NUREG–1412

Foundation for the Adequacy of the Licensing Bases

A Supplement to the Statement of Considerations
for the Rule on Nuclear Power Plant
License Renewal (10 CFR Part 54)

Final Report

Manuscript Completed: October 1991
Date Published: December 1991

Office of Nuclear Reactor Regulation
U.S. Nuclear Regulatory Commission
Washington, DC 20555

ABSTRACT

The objective of this report is to describe the regulatory processes that assures that any plant-specific licensing bases will provide reasonable assurance that the operation of nuclear power plants will not be inimical to the public health and safety to the end of the renewal period. It is on the adequacy of this process that the Commission has determined that a formal renewal licensing review against the full range of current safety requirements would not add significantly to safety and is not needed to assure that continued operation throughout the renewal term is not inimical to the public health and safety or common defense and security.

This document illustrates in general terms how the regulatory process has evolved in major safety issue areas. It also provides examples illustrating why it is unnecessary to re-review an operating plant's licensing basis, except for age-related degradation unique to license renewal, at the time of license renewal.

This report is a supplement to the Statement of Considerations for the Nuclear Regulatory Commission's rule (10 CFR Part 54) that establishes the criteria and standards governing nuclear power plant license renewal.

TABLE OF CONTENTS

1 INTRODUCTION

1.1 Objective of This Analysis

The objective of this analysis is to describe the regulatory processes that assures that the plant-specific licensing bases provide reasonable assurance that the operation of nuclear power plants will not be inimical to the public health and safety. It is because of the adequacy of this process that the Commission has determined that a formal renewal licensing review against the full range of current safety requirements would not add significantly to safety and is not needed to assure that continued operation throughout the renewal term is not inimical to the public health and safety.

This report supplements the Statement of Considerations for the Nuclear Regulatory Commission's rule (10 CFR Part 54) that establishes the criteria and standards governing the renewal of a nuclear power plant's operating license.

1.2 Scope of This Analysis

This document discusses in general terms how the regulatory process has evolved in areas of major safety concerns. It also provides examples illustrating why it is unnecessary to re-review an operating plant's licensing basis, except for age-related degradation unique to license renewal. Plant-specific details of how the regulatory processes have been implemented for specific technical areas can be found in the records of each plant's license applications and licenses. These are maintained in the docket files.

The statement of considerations for the license renewal rule includes an overview of the basis for the Commission conclusion that the regulatory processes provide reasonable assurance that the current licensing basis (CLB) provides an acceptable level of safety and that licensees comply with their CLB. The elements of the regulatory process relied on are identified and explained. Among these elements are the original licensing basis, the workings of the Commission's backfit policy, and the roles of operating event monitoring and safety issue resolution. In the present document, that overview is restated and is also supported by the addition of a substantial detailed examination of the CLB adequacy basis for the full range of specific areas of major safety issues. In the examinations for each of these areas, the safety issues involved are described, the key features of the regulatory requirements are noted, the evolution of the current licensing bases is explained, and conclusions are presented, stating once more the main foundations for the continued acceptability of the CLB for older as well as newer plants.

1.3 Technical and Policy Overview

1.3.1 Principles of the License Renewal

The license renewal rule rests on two key principles. The first principle is that, with the exception of age-related degradation unique to license renewal, the regulatory process provides reasonable assurance that the licensing bases of all currently operating plants provide and maintain an acceptable level of safety for operation during any renewal period. The second and equally important principle is that each plant's current licensing basis must be maintained throughout the renewal period, in part through a program of age-degradation management for systems, structures, and components that are important for license renewal. This report relates to the first of these principles that the regulatory process provides reasonable assurance of adequate protection to the public health and safety.

1.3.2 Current Licensing Basis

The current licensing basis (CLB) is the set of NRC requirements applicable to a specific plant and a licensee's written commitments for assuring compliance with and operation within applicable NRC requirements and the plant-specific design basis (including all modifications and additions to such commitments over the life of the license) that are docketed and in effect. The CLB includes the NRC regulations contained in 10 CFR Parts 2, 19, 20, 21, 30, 40, 50, 51, 54, 55, 70, 72, 73, 100, and appendices thereto; orders; license conditions, exemptions; and technical specifications. It also includes the plant-specific design-basis information defined in 10 CFR 50.2 as documented in the most recent final safety analysis report (FSAR) as required by 10 CFR 50.71 and the licensee's commitments remaining in effect that were made in such docketed licensing correspondence as such licensee responses to NRC bulletins, generic letters, and enforcement actions, as well as licensee commitments documented in NRC safety evaluations, or as described in licensee event reports.

The CLB differs among plants. These differences arise because plants are licensed at different times, at different sites, with different designs and individual operating experience. This document describes and discusses the regulatory processes designed to ensure that notwithstanding these differences the CLBs of all plants remain acceptable throughout plant life, including any renewal period. This document also notes the role of the backfit process in incorporating newly evolving requirements in previously licensed plants and the contribution made to safety by staff and industry guidance. Analyses in this document show in specific detail how these evolutionary processes have worked in the various safety-issue areas to provide reasonable assurance that the NRC-licensed nuclear power plants provides and maintains an acceptable level of safety.

1.3.3 Acceptable Level of Safety

The Atomic Energy Act of 1954, as amended, directs the Commission to ensure that nuclear power plants operation is not inimical to the health and safety of the public. However, this "not inimical" standard contemplated neither an absolute protection or zero risk and, therefore safety improvements beyond the minimum needed for adequate protection are possible. As new information is developed on technical subjects, the NRC reviews the potential safety concerns and then requires that plant designs be able to cope with the identified concern with sufficient safety margins and reliable systems. Should new information reveal an unforeseen significant safety concern or insufficient margins and backup capability, the new information is carefully evaluated and the Commission may, in light of the information, conclude that the existing regulations need to be changed in order to continue to assure an acceptable level of safety, or that some other regulatory action needs to be taken. Therefore, as the Commission identifies new issues or concerns, reasoned engineering decisions are made within the Commission about whether any additional measures must be taken at plants to resolve the issues. When specific actions are identified, the Commission, through its regulatory programs, can modify the licensing bases at operating plants at any time to resolve the new concern. The process of backfitting requirements to plants already licensed is currently guided by the provisions of the backfit rule (10 CFR 50.109). Before promulgation of the backfit rule, similar considerations were applied, although the backfit rule gave the process more structure.

1.3.4 Regulatory Oversight

The Commission's regulatory oversight programs ensure that the plant's licensing basis is modified as appropriate to reflect significant new information on technical topics, including the effects of age-related degradation affecting the design or operation of the licensed plant so that the licensing bases at operating plants continue to provide an acceptable level of safety. These continuing activities in place during the initial license term would continue to the end of the renewal term as well. Examples of such programs include operating events assessment and generic issues programs, discussed in the subsections that follow, as well as the Commission's inspection program. Should the Commission find that additional protection is needed to ensure the public health and safety or if significant additional protection at a reasonable cost would substantially enhance plant safety, the Commission may require the backfit of a licensed plant, that is, the addition of, elimination of, or modification to plant systems, structures, or components.

Operating Events Assessment

The Commission has an aggressive program for reviewing operating events at nuclear power plants. As a requirement of the current licensing basis, and one which would continue to the end of the renewal term, each licensee must notify the Commission promptly of any plant event that meets or exceeds the threshold defined in 10 CFR 50.72 and must file written licensee event reports (LERs) for those events that meet or exceed the threshold defined in 10 CFR 50.73. The staff reviews this information daily and follows up on events that appear to be potentially risk significant or are judged to be a possible precursor to a more severe event. Depending on the significance of the reported event, the staff may take further action to notify one or more classes of licensees or to impose additional requirements on some or all licensees. Industry groups, notably the Institute of Nuclear Power Operations (INPO), disseminate information in significant operating experience reports (SOERs). The total process offers a high degree of assurance that events which are potentially risk significant or are precursors to potentially significant events are being reviewed and resolved expeditiously.

Generic Issue Programs

As described in SECY-89-138, the Commission also maintains an active program for evaluating and resolving generic safety issues that may impact public health and safety. A generic safety issue (GSI) involves a safety concern that may affect the design, construction, or operation of all, several, or a class of reactors or facilities. Its resolution may have a potential for safety improvements and promulgation of new or revised requirements or guidance. The prioritization process, as described in NUREG-0933, evaluates the safety significance of an issue and classifies the issues as high, medium, or low priority GSIs. GSIs that are categorized as high priority are further evaluated to determined whether they involve questions regarding adequate protection of the public health and safety and therefore should be re-categorized as unresolved safety issues (USIs). GSIs are issues that involve enhancements to safety but do not call into question the adequacy of the current licensing basis. By contrast, USIs are defined as issues that potentially involve adequate protection of the public health and safety. Thus, a USI may represent a matter where the adequacy of the current licensing basis has not been established Resolution of a USI may result in a determination that action is necessary to ensure adequate protection, or it may result in a conclusion that, in fact, there are no concerns as to adequate protection of the public health and safety and further action is not warranted. The licensing basis of individual plants includes changes that have resulted from resolution of generic issues determined to be applicable and will include applicable generic-issue-derived changes in the future. It should be noted, however, that, as discussed immediately below, all currently unresolved GSIs address only enhancements of safety. This conclusion was determined during the initial evaluation of the generic concern, which assessed whether any aspect of the generic concern might have a significant impact on the protection of the public health and safety such that immediate action would be warranted. Since GSIs involve only enhancements to safety, implementation of the resolution of any GSI is not necessary to ensure public health and safety during the renewal term.

A special group of 22 GSIs deemed to be of sufficient significance to warrant both a high-priority resolution effort and special attention in tracking were designated as Unresolved Safety Issues (USIs). All USIs have been resolved. Most of the USI resolutions have been implemented; the remainder are being implemented on a satisfactory schedule. In one case, USI A-46, "Seismic Qualification of Equipment in the Operating Plants," the NRC and the utility groups are negotiating the implementation schedule in accordance with the NRC policy on integrated Schedule for Plant Modifications, Generic Letter 83-20, dated May 9, 1983.

This process for ensuring implementation of these remaining USIs is the same process used by the NRC in the past to ensure resolution and implementation of USIs. Furthermore, this process will be used in the future if the NRC identifies new issues that meet the definition of a USI.

The generic issues program is described and discussed more fully in Chapter 19.

1.3.5 Evolution of NRC Requirements

In the late 1960s and early 1970s, the U.S. Atomic Energy Commission's (now Nuclear Regulatory Commission) scope of review of proposed power reactor designs was evolving. In the same way that the scope of review evolved, the licensing bases for early plants have evolved as appropriate actions were identified through the NRC's continuing oversight activities and regulatory processes. In 1967, the Commission published for comment and interim use a proposed "General Design Criteria (GDC) for Nuclear Power Plants" that established minimum requirements for the principal design standards. The GDC were formally adopted in 1971 and have been used since that time as guidance in reviewing new plant applications. The general design criteria are contained in Appendix A to 10 CFR Part 50. They establish minimum broad requirements for the principal criteria for the materials, design, fabrication, testing, inspection, and certification of all structures, components, equipment, and systems in nuclear power plants that are important to safety. The staff's plant-specific reviews with respect to the various areas of safety (discussed in Chapters 2 through 19) must arrive at a conclusion that the overall plant design satisfies the intent of the GDC requirements and that the plant can safely be operated.

Safety guides issued in 1970 became part of the regulatory guide series in 1972. These guides described methods acceptable to the staff for implementing specific portions of the regulations, including certain GDC, and formalized staff techniques for reviewing a facility. In 1972, the Commission distributed for information and comment a proposed "Standard Format and Content of Safety Analysis Reports for Nuclear Power Plants," now Regulatory Guide 1.70. It gave applicants a standard format for these reports and identified the principal information needed by the staff for its review. The Standard Review Plan (SRP), NUREG-75/087, was published in December 1975 and revised in July 1981 (NUREG-0800) to give the staff more guidance for improving the quality and uniformity of its reviews. This guidance consisted of acceptance criteria and review procedures necessary to provide the staff with the basis for concluding that applicable GDC have been satisfied. For the most part, the detailed acceptance criteria prescribed in the SRP were not new; rather, they were methods of review that, in many cases, had not been previously published in any regulatory document.

1.3.6 Systematic Evaluation Program

In 1977 the NRC initiated the Systematic Evaluation Program (SEP) to review the designs of 10 of the oldest operating nuclear power plants and thereby confirm and document their safety. The reviews were organized into approximately 90 review topics (reduced by consolidations from 137 originally identified).

The SEP effort highlighted a group of 27 regulatory topics for which corrective action was generally found to be necessary for the initial SEP plants and for which safety improvements for other operating plants of the same vintage could be expected. The topics on this smaller list are referred to as the SEP lessons learned, and the Commission expects that these topics would be generally applicable to operating plants that received their construction permits in the late 1960s or early 1970s.

Of the 27 regulatory topics highlighted in the SEP effort, four have been completely resolved and one is of such low safety significance as to require no regulatory action. Of the 22 issues remaining open, the Commission has determined that none require immediate action to protect the public health and safety.

The Commission has incorporated the 22 issues into the established regulatory process for determining the safety importance of GSIs and has determined that none require immediate action as part of a license renewal application. As with the case for GSIs and USIs, the existing prioritization process to be used during the review and prioritization of the SEP lessons learned issues should prove to be adequate in the future to resolve these issues.

1.3.7 Review of Changes in NRC Requirements and Guidance

The NRC has arranged for systematic review to help ensure the effectiveness, efficiency, and coherence of changes in requirements and guidance. The Committee To Review Generic Requirements (CRGR) has the responsibility to review and recommend to the Executive Director for Operations (EDO) approval or disapproval of requirements or staff positions to be imposed by the NRC staff on one or more classes of power reactors. This review applies to staff proposals of requirements or positions that reduce existing requirements or positions and proposals that increase or change requirements. This responsibility is implemented in such a manner as to ensure that the provisions of 10 CFR 2.204, 10 CFR 50.109, and 10 CFR 50.54(f) pertaining to generic requirements and staff positions are carried out by the staff. The CRGR process aims to eliminate or remove any unnecessary burdens placed on licensees, reduce the exposure of workers to radiation in adhering to some of these requirements, and conserve NRC resources -- all to be done without impairing the adequate protection of the public health and safety and with furthering the review of new, cost-effective requirements and staff positions. The CRGR and the associated staff procedures are intended to ensure NRC staff implementation of 10 CFR 50.54(f) and 50.109 for generic backfit matters. By having the committee submit recommendations directly to the EDO, only a single agency-wide point of control is transmitted.

For those rare instances where it is judged that an immediately effective action is needed to ensure that facilities pose no undue risk to the health and safety of the public (10 CFR 50.109(a)(4)(ii)), no prior review by the CRGR is necessary. However, before or after any such action, the staff conducts a documented evaluation that includes a statement of the objectives of and reasons for the actions and the basis for invoking the exception.

In earlier years, the function of the CRGR to determine the need for backfit was performed by the Regulatory Requirements Review Committee; the CRGR process introduced enhanced discipline and documentation to the determination.

Throughout the process of evolution of requirements and staff guidance, the Commission has had the benefit of advice on significant safety issues from the Advisory Committee on Reactor Safeguards (ACRS).

1.3.8 Detailed Analyses

Chapters 2 through 19 of this report describe and discuss, in more specific terms, how the previously stated regulatory programs and processes have worked in major administrative, technical, and procedural areas, thereby detailing the Commission's reasons for considering it unnecessary to review an operating plant's licensing basis, except for age-related degradation unique to license renewal, at the time of license renewal. These discussions are indicative of how the regulatory process will continue to ensure that, despite the variations in plant design criteria, an operating reactor will continue to provide an acceptable level of safety.

1.4 Conclusions

The processes outlined above and discussed more substantially in the specific areas on major safety issues supports as the justification for focusing the NRC review to only age-related degradation unique to license renewal. The key elements include the original licensing basis and the Commission rules, regulations, requirements, and reviews underlying it; the Commission's backfit policy, which has historically resulted in backfit of new requirements when required for adequate protection of public health and safety, compliance with the plant licensing basis, or when cost-justified as safety enhancements; and the Commission's programs of inspection, of monitoring operational events, and for resolution of plant-specific and generic safety issues. It is through these processes and programs that the staff ensures the continued acceptability of the licensing bases of all plants.

2 SITE-RELATED ISSUES

The following section discusses a number of site-related topics that include general site geography and demography, potential site-vicinity hazards, and potential accidents from natural phenomena. The natural phenomena discussed in this section include regional climatology and local meteorology, site hydrology and potential flooding issues, as well as geology and related seismic considerations. The nature of the licensing basis for each of these topics will be discussed in greater detail below.

2.1 Geography, Demography, and Potential Site-Proximity Hazards

2.1.1 Scope

Site geography includes the consideration of site location and description, exclusion area authority and control, and broad land use patterns. Demography covers the population distribution in the vicinity of the site. With respect to both, geography and demography, changes over the licensing term are addressed by making conservative projections for land use and population changes.

Nearby industrial, transportation, and military facilities can pose a threat to the safe operation of a nuclear power plant. Site acceptability depends, in part, on the adequacy of the licensee's assessment of and protection against the hazards posed by nearby man-made activities. Although these activities present a wide range of external events, the hazards associated with them can be grouped into four broad categories: (1) missiles, (2) explosions, (3) fires, and (4) toxic gases.

2.1.2 Safety Issues and Regulatory Requirements

The Commission regulations require that site evaluation factors be considered in the review of a license application, including those relating to site location, exclusion area, low population zone, and population center distance. In addition, the regulations require that plant systems, structures, and components important to safety be appropriately protected against the dynamic effects from events and conditions outside the nuclear power unit.

2.1.3 Evolution of Current Licensing Basis

Prior to the issuance of the Standard Review Plan (SRP), the staff reviewed site exclusion area control and demography on a case-by-case basis, with emphasis on independent verification of site-specific characteristics such as property ownership, mineral rights, and nearby population distributions. It was recognized that land use and local and State regulations can and do change with time. Hence, for example, staff review of the licensee's ability to exercise appropriate authority and control of the activities within the exclusion area focused on those aspects that were deemed to be time-dependent.

With respect to demography, efforts were made to obtain reasonable projections of population distributions to the end of the licensing term. This was done in recognition that population changes, driven by such factors as local socio-economic conditions, can be potentially significant with passage of time.

Similarly, potential external hazards in the vicinity of the site that could affect plant safety were reviewed on a case-by-case basis, with emphasis on independent verification of individual hazards for each site. Review of numerous license applications in the early and middle 1970s led to a two-part approach in addressing site hazards. First, the site and its surroundings were examined with the intent of identifying all existing hazards that had a potentially significant impact on the safe operation of the plant. Once the hazards were identified, they were evaluated in terms of their severity and likelihood.

For hazardous industrial or military activities, conservative projections were made concerning transportation traffic or accident rates. Where projections were not feasible, conservatism commensurate with the potential changes to the hazardous materials operations was applied to the current conditions.

The SRP was published in the early 1970s to improve the quality and uniformity of staff review in the particular subject areas as well as to specify acceptance criteria of the staff for concluding that the applicable regulations had been satisfied. The detailed acceptance criteria contained in the SRP were not new, but rather standardized the acceptance practices which had been established in previous reviews.

As mentioned earlier, the regulations require that reactors be protected against the dynamic effects of events and conditions outside the nuclear plant. For external hazards such as toxic gases or airplane hazards, the licensees frequently provided information, during the initial licensing review, concerning the frequency of shipments or the amount of airplane traffic and then proposed plant protection features or determined that no protection was necessary based on a determination or analysis of the hazard. Because the regulations remain in effect for the term of the license, licensees have the responsibility to ensure that the plant remains appropriately protected from any site-related hazards, new or existing at the time the plant was licensed. The staff recognizes that licensees cannot control development around the site and the regulations do not require licensees to do so. However, the regulations clearly place the responsibility on licensees for ensuring the protection of the reactor.

The Commission inspection activities also provide another source of information concerning changes that are occurring around the reactor site. The resident inspector, who typically resides in the area of the plant, has direct knowledge and access to the local media and therefore can be informed of potential developments in the surrounding environment that can potentially affect plant safety. In addition, regional and headquarters-based inspectors routinely visit sites, thereby affording further opportunities for observations of potential changes of the surrounding environment. As new issues are identified, these issues are raised to both the licensee and the staff for resolution. The staff is undertaking revisions to selected inspection procedures to require routine documentation of potential changes in the general environs of the facility.

2.1.4 Conclusions

The staff review at the time of the initial licensing of a facility determined that the information provided by the applicant was sufficient for the staff to conclude that the plant siting met the staff acceptance criteria and the intent of all applicable regulations. The staff has and will continue to obtain new information related to this subject area through a variety of sources such as updates to the FSAR, research reports, operating plant events and routine plant inspections. The staff reviews this information and, in the past, has required licensees to take actions to upgrade the plant to provide continuing assurance of adequate protection of the public health and safety. In addition, the staff will continue to review new information in this subject area and if the staff determines that new or different requirements are needed, the staff has the capability within the existing regulatory process to require additional analyses or plant modifications, as necessary, to ensure the continued health and safety of the public. In conclusion, the Commission concludes that the current regulatory process has and will continue to provide reasonable assurance that the licensing basis of all currently operating plants are sufficient to assure that operation is not inimical to the public health and safety.

2.2 Meteorology

2.2.1 Scope

This section discusses the regulations and licensing requirements used by the NRC to ensure safe siting and operation of nuclear power plants with respect to meteorology. Site-specific data on meteorology and regional climatology is used to ensure that continuing staff awareness of both regional and local meteorological trends is maintained to ensure that the design basis meteorology conditions remain sufficient to ensure safe plant operation in accordance with the current licensing basis (CLB).

2.2.2 Safety Issues and Regulatory Requirements

Nuclear power plants are designed, operated, and maintained such that offsite exposure to accidental gaseous releases and their resultant dose to receptors at the plant exclusion area boundary, low population zone distance, and the population center and offsite exposures from routine normal operational releases at nearby receptors including residences, dairies, farms, schools, etc., comply with the requirements of 10 CFR Part 100 and 10 CFR Part 50, Appendix I, respectively.

2.2.3 Evolution of Current Licensing Basis

"Meteorology and Atomic Energy-1968," AEC 1968, generally, served as a compendium of meteorology measurement, dispersion modeling, and dose determination during the early 1970s. This document was later used as the foundation for issuance of safety guides, regulatory guides, and the Standard Review Plan sections in the meteorology area of review.

Regarding the regional climatology, the long-term meteorological conditions affecting the plant vicinity, as described in the plant final safety analysis report (FSAR) submitted by the licensee, and addressed in the staff safety evaluation report (SER), are not expected to change significantly. The basis for the expectation of minimal change lies in the meteorological and climatological records collected from as far back as 1895 and published in documents prepared by the National Climatic Center. The published data demonstrate that nearly constant climatological conditions exist in a local area with only small anomalies on a monthly or seasonal basis. Studies by the National Institute of Standards and Technology (formerly National Bureau of Standards) using long-term data show the low probabilities of rapid significant changes in long-term temperature, precipitation, and wind speeds. Thus, design basis characteristics might only be exceeded for brief time periods in the near term (say, next 50 to 100 years), while greater frequency of duration of "abnormal" conditions might provide evidence of a possible worldwide major climatic change. The changes of global extent and recognition of its happening would be expected to provide sufficient time to allow the staff to take appropriate measures to deal with the changes such that the current licensing basis continues to remain valid for the licensed term.

Recently, the staff considered the changes to global climate due to atmospheric ozone depletion in response to concerns raised by the Council on Environmental Quality (CEQ) pursuant to NEPA. The thrust of the memorandum dealt with high- and low-level waste repositories and impact of global warming on rainfall and flooding due to changes in climate. The original plant site evaluation was based on the premise and expectation that site area climate would generally not change significantly during the operational life of the plant. Based on current information and staff review experience, site data concerning regional climate provided in FSARs and staff conclusions contained in plant SERs are expected to remain valid during any extended term of the operating license.

With respect to the review area of local meteorology, the local meteorology conditions, generally, reflect the expected regional climate influence and unique topographic features that may result in micro-scale phenomena that had been addressed in the FSAR and SER and should remain the same as described in those documents during any renewal period. However, in the event of a marked climatic change, it is probable that parameters representing "normals," such as extreme wind, precipitation, temperature, and structural capacities may require reevaluation, since climatic change may result in storm systems with greater intensities and frequency than those assumed in the design basis of the plant. However, this type of reevaluation is a part of the NRC staff's continuous plant safety assessment effort to ensure continued adequacy of the meteorology-related current licensing basis for operating plants. The licensee routinely publishes specific information related to meteorology in semiannual reports of meteorological joint frequency data as required by emergency planning requirements. Plant modifications or improvements in the meteorological monitoring system dictated by the semiannual reports have been implemented by the staff or licensee to ensure the validity of the current licensing basis.

The current licensing basis requires each plant to have an onsite meteorological monitoring program. This onsite meteorological measurement program continues to monitor local conditions that would affect the dispersion of radioactive and toxic gaseous effluent from or to the plant.

2.2.4 Conclusions

The staff review at the time of the initial licensing of a facility determined that the information provided by the applicant was sufficient for the staff to conclude that the meteorological and climatological factors related to the siting of the facility met the staff acceptance criteria and the intent of all applicable regulations. The acceptance criteria are constructively established to ensure adequate protrection from extremely low probability meteorological phenomena. The design basis for these extremely low probability meteorological events is related with conservation and is not experted to change appreciably during the next 50 to 100 year period. However, the staff has and will continue to obtain new information related to this subject area through a variety of sources such as updates to the FSAR, research reports, operating plant events and routine plant inspections. The staff reviews this information and, in the past, has required licensees to take actions to upgrade the plant to provide continuing assurance of adequate protection of the public health and safety. In addition, the staff will continue to review new information in this subject area and if the staff determines that new or different requirements are needed, the staff has the capability within the existing regulatory process to require additional analyses or plant modifications, as necessary, to ensure the continued health and safety of the public. In conclusion, the Commission concludes that the current regulatory process has and will continue to provide reasonable assurance that the licensing basis of all currently operating plants are sufficient to assure that operation is not inimical to the public health and safety.

2.3 Hydrologic Engineering

2.3.1 Scope

Nuclear power plants interact continuously with their hydrosphere (e.g., rivers, lakes, coastal environments, ground water, and water control structures). Such interactions present potential hazards of flooding as a result of severe hydro-meteorological conditions. This section discusses the regulations and licensing requirements adopted by the NRC to ensure safe operation of nuclear power plants against severe flooding hazard over the licensed plant life. The continuous review process used by the NRC staff to assess safety impact resulting from changes in hydrometeorological parameters and new information related to flooding hazard is also discussed. Lastly, the rationale for an NRC staff conclusion that the current licensing basis for operating plants is adequate to protect public health and safety is discussed.

2.3.2 Safety Issues and Regulatory Requirements

The Commission's regulations require, in part, that systems, structures, and components important to safety be designed to withstand the effects of natural phenomena such as floods, tsunami, and seiches without loss of capability to perform their safety functions. In addition, the regulations require that physical characteristics of the site, including meteorology and hydrology, be taken into account in determining the acceptability of a site for a nuclear power reactor. More specifically, the regulations require that a detailed study be performed and that the design bases for seismically induced floods and water waves be based on the results of the required geologic and seismic investigations and that these design bases be taken into account in the design of the nuclear power plant.

In order to demonstrate compliance with the above regulatory requirements, nuclear power plants are designed to prevent or mitigate the loss of capability for cold shutdown and maintenance thereof resulting from the most severe flood conditions that can reasonably be predicted to occur at a site as a result of severe hydrometeorological conditions, seismic activity, or both.

2-5

2.3.3 Evolution of Current Licensing Basis

The NRC has recognized the potential hazards resulting from the flooding of a commercial nuclear power plant since the mid-1960s. The flooding hazard review of plants licensed in the late 1960s used a formal and disciplined evaluation process but on a plant-specific basis.

In order to independently evaluate the potential for flooding at proposed reactor sites, the Atomic Energy Commission contracted the U.S. Army Corps of Engineers and the U.S. Geological Survey to evaluate flooding potential at coastal sites and river sites, respectively. In 1970 the AEC staff developed specific guidance for use in determining flood protection requirements for all plant sites. The staff adopted the concept of the "Probable Maximum Flood" from the Corps of Engineers and applied this concept to sites along streams and rivers. Guidance for determining the Probable Maximum Hurricane Surge, Probable Maximum Seiche, and Probable Maximum Tsunami flooding was also developed and applied for plant sites along lakes and oceans.

During 1973 through 1975, the NRC, based on the above work, published integrated staff positions and acceptance criteria related to acceptable design of nuclear power plants against flooding hazard.

As part of the Commission's process of reviewing new information related to specific technical issues, the staff learned that a number of probabilistic risk assessments completed between 1981 and 1987 concluded that external flooding could be a key contributor to overall plant risk. As a result, the NRC staff determined that it would be useful to evaluate the continued adequacy of the existing flood protection regulations by performing an individual plant examination at each reactor site. The staff developed regulatory guidance and acceptance criteria for the evaluation of plant-specific vulnerabilities to beyond design bases events initiated from a severe flooding event and incorporated this evaluation into the Individual Plant Examination External Events (IPEEE) program. The staff intends to evaluate the results of the IPEEE program, not on a plant-specific basis, but as an aggregate to determine appropriate changes to the regulations related to protections from potential external flooding. If any changes are identified, modification of the regulations will proceed and implementation of any plant modifications would be required on plant-specific bases to meet the revised regulations.

In October 1989, NRC issued Generic Letter 89-22 to inform licensees that the staff has adopted for future plants the latest probable maximum precipitation (PMP) criteria published by the National Oceanic and Atmospheric Administration (NOAA), National Weather Services (NWS), to establish acceptable design configurations for safety-related nuclear power plant facilities. In this letter the staff also concluded and stated that the existing criteria for determining PMP at operating plants still provided reasonable assurance of the protection of the health and safety of the public and that no additional backfit action by licensees was necessary.

2.3.4 Conclusions

The staff review at the time of the initial licensing of a facility determined that the information provided by the applicant was sufficient for the staff to conclude that the design to protect the facility from floods met the staff acceptance criteria and the intent of all applicable regulations. The staff has and will continue to obtain new information related to this subject area through a variety of sources such as updates to the FSAR, research reports, operating plant events and routine plant inspections. The staff reviews this information and, in the past, has required licensees to take actions to upgrade the plant to provide continuing assurance of adequate protection of the public health and safety. In addition, the staff will continue to review new information in this subject area and if the staff determines that new or different requirements are needed, the staff has the capability within the existing regulatory process to require additional analyses or plant modifications, as necessary, to ensure the continued health and safety of the public. In conclusion, the Commission concludes that the current regulatory process has and will continue to provide reasonable assurance that the licensing basis of all currently operating plants are sufficient to assure that operation is not inimical to the public health and safety.

2.4 Geologic, Seismologic, and Geotechnical Engineering

2.4.1 Scope

This section discusses the regulations and licensing requirements adopted by the NRC to ensure safe operation of nuclear plants subject to the influence of site-specific geologic, seismologic, and geotechnical hazards over the licensed plant life. The ongoing review process used by NRC staff to assess the safety impact resulting from changes in parameters related to the hazards and pertinent new information is also discussed.

2.4.2 Safety Issues and Regulatory Requirements

The Commission's regulations require that systems, structures, and components important to safety be designed, fabricated, erected, and tested to quality standards commensurate with the importance of the safety function to be performed and that nuclear power plant systems, structures, and components important to safety be designed to withstand the effects of natural phenomena such as earthquakes without loss of capability to perform their safety functions. In addition, the regulations require, in part, that suitable redundancy be provided for the cooling water system to ensure that its safety function can be accomplished and that measures be established to ensure design control, material control, special processes control, and inspection and test controls. The regulations also require that all nuclear power plants be designed so that, if the safe shutdown earthquake (SSE) occurs, all safety-related systems, structures, and components remain functional.

In order to demonstrate compliance with these regulatory requirements, nuclear power plants are designed to prevent or mitigate the loss of capability for cold shutdown and maintenance thereof resulting from the safe shutdown earthquake, foundation settlement, or instability.

2.4.3 Evolution of Current Licensing Basis

When the NRC first began to review nuclear power plant seismic designs in the 1960s, the reviews were performed using a formal evaluation process but on a plant-specific approach.

In 1971, the General Design Criteria (GDC) for Nuclear Power Plants were formally adopted as the minimum requirements for the principal design standards. These requirements generally adopted the existing staff practice in effect at that time. The GDC have been used as guidance in reviewing new plant applications since then.

Because of the evolutionary nature of licensing requirements and the development of technology over the years, nuclear power plants employ a broad spectrum of design features and requirements depending on when the plant was designed and constructed, who was the manufacturer, and when the plant was licensed for operation.

The NRC staff, as part of its routine review process, continues to assess the potential safety impact of any new information related to geologic and seismologic issues. New information can be derived from research or additional data observations since the issuance of operating licenses. Whenever the results of such information indicated the need for remedial actions, including plant modifications, the staff has acted to ensure the implementation of such actions to ensure that the current licensing basis at potentially affected plants remains adequate to protect the health and safety of the public.

Examples of this include the discovery of the capable Hosgri fault near Diablo Canyon and the assessment of earthquakes occurring near Maine Yankee. One nuclear power plant (Humboldt Bay) and one non-power reactor (General Electric Test Reactor) were shut down, and remain permanently shut down, as a direct result of geologic concerns. Another example is the "Charleston Earthquake Issue" which was raised as a result of a U.S. Geologic Survey letter in 1982. This letter highlighted the possibility that large damaging earthquakes have some likelihood of occurring at locations not formally considered in past licensing decisions. The staff initiated the Seismic Hazard Characterization Project, which provided probabilistic seismic hazard estimates for nuclear power plant sites east of the Rocky Mountains. A similar project was carried out by the Electric Power Research Institute (EPRI) for the electric utility industry. The staff's purpose in evaluating the probabilistic studies has been to identify plants in the central and eastern United States where past licensing decisions have resulted in the potential for plant-specific vulnerabilities to beyond design basis events with respect to seismic hazard. The staff's plan for documenting and reconfirming the degree of protection from seismic safety issues is part of the staff resolution of Generic Issue A-46, "Seismic Qualification of Equipment in Operating Nuclear Power Plants."

The purpose of the A-46 is to reverify and document the seismic adequacy of mechanical and electrical equipment qualification to ensure the survival and functionality of equipment required to safely bring the reactor and plant to a safe shutdown condition. Consistent with the guidance for developing an unresolved safety issue, the staff performed an analysis to determine if the identified seismic concern might have a significant impact on the protection of the public health and safety and, therefore, that immediate remedial action would be warranted. The staff's conclusions and their technical bases have been published in NUREG-1211, "Regulatory Analysis for Resolution of Unresolved Safety Issue A-46, Seismic Qualification of Equipment in Operating Plants," February 1987. In this document, the staff concluded that equipment installed in nuclear power plants is inherently rugged and not susceptible to seismic damage. However, the staff also concluded that sufficient justification of a safety benefit could be made because, although the equipment was inherently rugged and not susceptible to seismic damage, failures resulting from seismic loads were possible if the equipment was not adequately supported or anchored. As a result, the staff issued Generic Letter 87-02 to all operating reactors, which required, as a backfit under 50.109, that licensees reverify the seismic qualification and anchorage of installed equipment to provide additional assurance of the continued protection of public health and safety.

In the staff review of areas related to plant foundation stability/settlement, water control structural safety and heat sink integrity, etc., the staff has generally upgraded established geotechnical engineering criteria and methodologies, which have been widely used to ensure full compliance with the necessary regulatory requirements.

As part of the staff's routine review effort, any new geotechnical engineering data or analysis techniques, which are judged by the staff as pertinent for inclusion in the current licensing criteria, have been incorporated following established NRC procedures. Where appropriate, the new analysis techniques were applied to assess design adequacy and any plant-specific modifications were implemented on plant-specific bases in order to ensure continued validity of the plant-specific licensing basis. Examples of such staff actions are (1) resolution of Waterford basemat cracking and structural integrity issue, (2) resolution of North Anna buried piping settlement and support integrity issue, and (3) resolution of San Onofre Unit 1 settlement of foundation and buried equipment issue.

As part of the staff's effort to assess the continued adequacy of the existing regulations, the staff has implemented the Individual Plant Examination External Events program. As part of this program, licensees will be requested to look for potential plant-specific vulnerabilities to beyond design basis accidents initiated from postulated seismic events and to report their findings to the Commission. The staff intends to use the results from all the plants in aggregate to determine if deficiencies exist in present regulations governing seismic hazards. If such deficiencies are identified, the staff intends to modify the regulations as necessary and would require plant-specific modifications as necessary to establish compliance with the new regulations.

2.4.4 Conclusions

The staff review at the time of the initial licensing of a facility determined that the information provided by the applicant was sufficient for the staff to conclude that the seismic design of the facility met the staff acceptance criteria and the intent of all applicable regulations. The staff has and will continue to obtain new information related to this subject area through a variety of sources such as updates to the FSAR, research reports, operating plant events and routine plant inspections. The staff reviews this information and, in the past, has required licensees to take actions to upgrade the plant to provide continuing assurance of adequate protection of the public health and safety. For example, a program currently in process is requiring licensees to confirm and document the seismic qualification and anchorage of equipment and, where necessary, to make plant specific modifications to ensure the continued adequate protection of public health and safety. As part of the continuing assessment of the adequacy of the existing regulations, the Commission has requested licensees to evaluate their plants for potential vulnerabilities for beyond design basis accidents resulting from geologic, seismologic, and geotechnical hazard considerations. If the results of this evaluation indicate that the existing regulations need modification, then the staff will proceed to revise the regulations and require plants to meet the revised regulations. In addition, the staff will continue to review new information in this subject area and if the staff determines that new or different requirements are needed, the staff has the capability within the existing regulatory process to require additional analyses or plant modifications, as necessary, to ensure the continued health and safety of the public. In conclusion, the Commission concludes that the current regulatory process has and will continue to provide reasonable assurance that the licensing basis of all currently operating plants are sufficient to assure that operation is not inimical to the public health and safety.

3 DESIGN OF STRUCTURES, COMPONENTS, EQUIPMENT, AND SYSTEMS

3.1 Scope

This section addresses the principal criteria required for the materials, design, fabrication, testing, inspection, and certification of all structures, components, equipment, and systems that are important to safety. Important to safety is defined in the introduction to 10 CFR Part 50, Appendix A, as those systems, structures, and components that provide reasonable assurance that the facility can be operated without undue risk to the health and safety of the public. The topics covered by this section are: the NRC General Design Criteria; classification of systems, structures, and components; wind and tornado loadings and water level (flood) design; missile protection; protection against dynamic effects associated with postulated rupture of piping; seismic design; design of Seismic Category I structures, mechanical systems, and components; seismic and dynamic qualification of Seismic Category I mechanical and electrical equipment; and environmental design of mechanical and electrical equipment.

3.2 Conformance With NRC General Design Criteria

3.2.1 Safety Issues and Regulatory Requirements

10 CFR Part 50, Appendix A, contains General Design Criteria (GDC) that establish minimum broad requirements for the principal criteria mentioned in Subsection 3.1 above.

3.2.2 Evolution of Current Licensing Basis

In the late 1960s and early 1970s, the U.S. Atomic Energy Commission's (now Nuclear Regulatory Commission) scope of review of proposed power reactor designs was evolving and somewhat less defined than it is today. The requirements for acceptability evolved as new facilities were reviewed. In 1967, the Commission published for comment and interim use proposed General Design Criteria (GDC) for Nuclear Power Plants that established minimum requirements for the principal design standards. The GDC were formally adopted in 1971 and have been used as guidance in reviewing new plant applications since that time. Safety guides issued in 1970 became part of the Regulatory Guide Series in 1972. These guides describe methods acceptable to the staff for implementing specific portions of the regulations, including certain GDC, and formalize staff techniques for performing a facility review. In 1972, the Commission distributed for information and comment a proposed "Standard Format and Content of Safety Analysis Reports for Nuclear Power Plants," now Regulatory Guide 1.70. It provided a standard format for these reports and identified the principal information needed by the staff for its review. The Standard Review Plan (SRP), NUREG-75/087, was published in December 1975 and updated in July 1981 (NUREG-0800) to provide further guidance for improving the quality and uniformity of staff reviews. This guidance consisted of acceptance criteria and review procedures necessary to provide the staff with the basis for concluding that applicable GDC have been satisfied. For the most part, the detailed acceptance criteria prescribed in the SRP were not new; rather they were methods of review that, in many cases, were not previously published in any regulatory document.

Because of the evolutionary nature of the licensing requirements discussed above and the developments in technology over the years, operating nuclear power plants embody a broad spectrum of design features and requirements depending on when the plant was constructed, who was the manufacturer, and when the plant was licensed for operation. The amount of documentation that defines these safety-design characteristics also has changed with the age of the plant. Although the earlier safety evaluations of operating facilities did not address many of the topics discussed in current safety evaluations, all operating facilities have been reviewed more recently against a substantial number of major safety issues that have evolved since the operating license was issued. Conclusions of overall adequacy with respect to these major issues (e.g., emergency core cooling system, fuel design and pressure vessel design) are a matter of record. On the other hand, a number of other issues (e.g., seismic considerations, tornado and turbine missiles, flood protection, pipe break effects inside containment, and pipe whip) were not originally reviewed against today's acceptance criteria for many operating plants.

The Systematic Evaluation Program (SEP) was initiated by the staff in 1977 to review the designs of older operating nuclear power plants in order to enhance the documentation of their safety. The review provided (1) an assessment of the significance of differences between current technical positions on safety issues and those that existed when a particular plant was licensed, (2) a basis for deciding on how these differences should be resolved in an integrated plant review, and (3) a documented evaluation of plant safety. The results of the staff's SEP reviews are documented in a series of Integrated Plant Safety Assessment Reports.

3.2.3 Conclusions

The staff review at the time of the initial licensing of a facility determined that the information provided by the applicant was sufficient for the staff to conclude that the plant design met the staff acceptance criteria, and the intent of all applicable regulations. The staff has and will continue to obtain new information related to this subject area through a variety of sources such as updates to the FSAR, research reports, operating plant events and routine plant inspections. The staff reviews this information and, in the past, has required licensees to take actions to upgrade the plant to provide continuing assurance of adequate protection of the public health and safety. In addition, the staff will continue to review new information in this subject area and if the staff determines that new or different requirements are needed, the staff has the capability within the existing regulatory process to require additional analyses or plant modifications, as necessary, to ensure the continued health and safety of the public. In conclusion, the Commission concludes that the current regulatory process has and will continue to provide reasonable assurance that the licensing basis of all currently operating plants are sufficient to assure that operation is not inimical to the public health and safety.

3.3 Classification of Structures, Components, and Systems

3.3.1 Safety Issues and Regulatory Requirements

The Commission regulations require that systems, structures, and components important to safety be designed, fabricated, erected, and tested to quality standards commensurate with the importance of the safety functions to be performed. Regulatory Guide 1.26, "Quality Group Classifications and Standards for Water-, Steam-, and Radioactive-Waste-Containing Components of Nuclear Power Plants," contains staff guidance that may be used to determine how a plant-specific design satisfies the regulations.

In addition, the regulations require that systems, structures, and components important to safety be designed to withstand the effects of earthquakes without loss of capability to perform their safety functions. Regulatory Guide 1.29, "Seismic Design Classification," contains staff guidance that may be used to determine how a plant-specific design satisfies the applicable regulations by identifying all systems, structures, and components that should be classified as Seismic Category I.

3.3.2 Evolution of Current Licensing Basis

During the 1960s, when General Design Criteria and the ASME Section III Code for Nuclear Power Plant Components were evolving, the staff reviews of quality group and seismic classification were performed on a plant-specific basis. Decisions on the classifications for systems, structures, and components were made based on staff positions at that time relative to the importance to safety of each item. These positions were first documented in Safety Guides 26 and 29 in 1970 and 1971 and later in Regulatory Guides 1.26 and 1.29 in 1972.

Subsequent to an exchange of correspondence between the General Electric Company and the staff during 1973 and 1974, the staff developed its position on classifications of the main steam and feedwater lines for BWR/6 plants. This position allowed BWR/6 applicants the option of installing a third shutoff valve and a seismic restraint downstream of the outside isolation valve in these lines. If the applicant chose this option, then the main steam and feedwater lines could be classified as Quality Group D (Non-Nuclear Safety) and non-Seismic Category I downstream of the seismic restraint. This position was implemented by the staff during its reviews of all BWR/6 plants between 1975 and 1981.

3.3.3 Conclusions

The staff review at the time of the initial licensing of a facility determined that the information provided by the applicant was sufficient for the staff to conclude that the the quality group and seismic classification met the staff acceptance criteria and the intent of all applicable regulations. The staff has and will continue to obtain new information related to this subject area through a variety of sources such as updates to the FSAR, research reports, operating plant events and routine plant inspections. The staff reviews this information and, in the past, has required licensees to take actions to upgrade the plant to provide continuing assurance of adequate protection of the public health and safety. In addition, the staff will continue to review new information in this subject area and if the staff determines that new or different requirements are needed, the staff has the capability within the existing regulatory process to require additional analyses or plant modifications, as necessary, to ensure the continued health and safety of the public. In conclusion, the Commission concludes that the current regulatory process has and will continue to provide reasonable assurance that the licensing basis of all currently operating plants are sufficient to assure that operation is not inimical to the public health and safety.

3.4 Wind, Tornado, and Flood Protection

3.4.1 Safety Issues and Regulatory Requirements

Commission regulations require, in part, that systems, structures, and components important to safety shall be designed to withstand the effects of natural phenomena such as tornadoes, hurricanes, and floods.

3.4.2 Evolution of Current Licensing Basis

The Commission has recognized the potential hazards resulting from wind, tornadoes, and floods imposed on a nuclear power plant since the early 1960s. These early reviews were performed on a plant-specific basis and the staff's guidelines for acceptability evolved as new facilities were reviewed. In 1975, the staff published sections of the Standard Review Plan addressing these subject areas. These guidance documents formally published regulatory positions that had been in practice at that time but had not been previously published in any type of regulatory document. Minor revisions were made to these SRPs in 1981; however, guidelines therein have remained virtually unchanged since 1975.

3.4.3 Conclusions

The staff review at the time of the initial licensing of a facility determined that the information provided by the applicant was sufficient for the staff to conclude that the designs for protection against tornadoes, and floods met the staff acceptance criteria and the intent of all applicable regulations. The staff has and will continue to obtain new information related to this subject area through a variety of sources such as updates to the FSAR, research reports, operating plant events and routine plant inspections. The staff reviews this information and, in the past, has required licensees to take actions to upgrade the plant to provide continuing assurance of adequate protection of the public health and safety. In addition, the staff will continue to review new information in this subject area and if the staff determines that new or different requirements are needed, the staff has the capability within the existing regulatory process to require additional analyses or plant modifications, as necessary, to ensure the continued health and safety of the public. In conclusion, the Commission concludes that the current regulatory process has and will continue to provide reasonable assurance that the licensing basis of all currently operating plants are sufficient to assure that operation is not inimical to the public health and safety.

3.5 Missile Protection

3.5.1 Safety Issues and Regulatory Requirements

Although large steam turbines and their auxiliaries are not safety-related systems as defined by NRC regulations, failures that occur in these turbines can produce large, high-energy missiles. If such missiles were to strike and to damage plant safety-related systems, structures, and components, they could render them unavailable to perform their safety function. Consequently, the regulations require, in part, that systems, structures, and components important to safety be appropriately protected against the effects of missiles that might result from such failures.

3.5.2 Evolution of Current Licensing Basis

The standard review plans for turbine missile protection were first published in NUREG-75/087 in December 1975 and updated in NUREG-0800 in July 1981. Regulatory Guide 1.115 was revised in July 1977. The staff's guidance in these documents indicates that the hazard rate for the loss of essential safety systems from a single turbine missile event must be less than 10^{-7} per year. Plants constructed prior to the publication of the staff guidance and evaluated as part of the Systematic Evaluation Program (SEP) were reviewed to ensure that an adequate level of protection from turbine missiles existed at these plants. Turbine missiles are identified as SEP Topic III-4.B.

In the initial reviews of this topic, the value for the probability of turbine failure resulting in ejection of turbine fragments through the turbine casing (P_1) was assumed be a constant of 10^{-4} per year for all turbines. Licensees or applicants evaluated the strike probability (P_2) and the damage probability (P_3) to ensure that the product of P_1, P_2, and P_3 is less than 10^{-7} per year. These early reviews indicated that large uncertainties exist in P_2 and P_3 because of the difficulty in modeling missiles, barriers, obstacles, and trajectories and in determining critical impact energies. In an Electric Power Research Institute sponsored seminar on "Turbine Missile Effects in Nuclear Power Plants" in October 1982, the staff indicated that, because of the uncertainties in P_2 and P_3, they would emphasize the missile generation probability (P_1) in future turbine missile reviews. P_2 and P_3 probabilities are to be order-of-magnitude estimates that are dependent on the orientation of the turbine to essential safety systems. The revised method also ensures that the hazard rate would be less than 10^{-7} per year. In this method the staff evaluates the procedures and methods used by turbine manufacturers to calculate the total missile generation probability and the associated turbine maintenance and inspection procedures.

3.5.3 Conclusions

The staff review at the time of the initial licensing of a facility determined that the information provided by the applicant was sufficient for the staff to conclude that the protection from turbine missiles met the staff acceptance criteria and the intent of all applicable regulations. The staff has and will continue to obtain new information related to this subject area through a variety of sources such as updates to the FSAR, research reports, operating plant events and routine plant inspections. The staff reviews this information and, in the past, has required licensees to take actions to upgrade the plant to provide continuing assurance of adequate protection of the public health and safety. In addition, the staff will continue to review new information in this subject area and if the staff determines that new or different requirements are needed, the staff has the capability within the existing regulatory process to require additional analyses or plant modifications, as necessary, to ensure the continued health and safety of the public. In conclusion, the Commission concludes that the current regulatory process has and will continue to provide reasonable assurance that the licensing basis of all currently operating plants are sufficient to assure that operation is not inimical to the public health and safety.

3.6 Protection Against Dynamic Effects Associated With Postulated Rupture of Piping

3.6.1 Safety Issues and Regulatory Requirements

Commission regulations require that systems, structures, and components important to safety be appropriately protected against the dynamic effects that may result from equipment failures, including the effects of pipe whipping and discharging fluids.

3.6.2 Evolution of Current Licensing Basis

In 1972, the staff documented the deterministic criteria that the staff had been using for several years as guidelines for selecting the locations and orientations of postulated pipe breaks inside containment and for identifying the measures that should be taken to protect safety-related systems and equipment from the dynamic effects of such breaks. Prior to use of these deterministic criteria, the staff used non-deterministic guidelines on a plant-specific basis. The staff criteria were subsequently revised and issued in May 1973 as Regulatory Guide 1.46, "Protection Against Pipe Whip Inside Containment."

Prior to 1972, the staff did not require postulation of pipe breaks outside containment. However, as a result of the continuing review of plant safety during that time period, the staff determined that such breaks should be postulated and the effects of these breaks should be evaluated by all licensees of operating plants and all applicants for Construction Permits or Operating Licenses. Therefore, generic letters were sent to all licensees and applicants from late 1972 through mid-1973. These letters provided deterministic criteria to be used for postulating pipe breaks outside containment and guidelines for evaluating the dynamic effects of these breaks. The letters requested that all recipients submit a report to the staff that summarized each plant-specific analysis of this issue. All operating reactor licensees and license applicants submitted the requested analyses in separate correspondence or updated the safety analysis report for the proposed plant to include the analysis. The staff reviewed all of these submitted analyses and prepared safety evaluations for all plants.

In November 1975, the staff published Standard Review Plan sections that slightly revised the two generic letters discussed above. As a part of its plant-specific reviews between 1975 and 1981, the staff used the guidelines in Regulatory Guide 1.46 for postulated pipe breaks inside containment and SRPs 3.6.1 and 3.6.2 for outside containment. In July 1981, SRPs 3.6.1 and 3.6.2 were revised to be applicable to both outside and inside containment. On June 19, 1987, Generic Letter 87-11 was issued to provide revised guidelines for locations of postulated pipe ruptures.

Another example of the continuing review of new technical issues is the potential problem of asymmetric loading on reactor vessel supports following a postulated reactor coolant pipe rupture. In 1975, the staff was informed that asymmetric loading on the reactor vessel supports resulting from a postulated reactor coolant pipe rupture at the vessel nozzle had not been considered in the original design of PWR plants. Following a brief review of this problem, the staff determined that a reevaluation of the reactor coolant system of all PWR plants was necessary to determine its capability to withstand these new loads. Unresolved Safety Issue (USI) A-2 was originated to address this problem. All licensees of PWR operating plants were requested to submit plant-specific analyses. In response to this request, several licensees formed an owners group and, in lieu of an analysis, submitted a report that incorporated advanced fracture mechanics techniques to demonstrate that a full diameter break could not occur in their primary loop piping. In Generic Letter 84-04, dated February 1, 1984, the staff agreed that such a break was unlikely to occur, provided it could be demonstrated by deterministic fracture mechanics analyses that postulated through-wall flaws in plant-specific piping would be detected by the plant's leakage monitoring systems long before the flaws could grow to unstable sizes. The concept underlying such analyses is referred to as "leak-before-break" (LBB). Subsequent evaluations of this issue by the staff led to the so-called "broad scope rule," which revised GDC-4 in 1987 to permit the use of LBB-type analyses in both PWRs and BWRs.

3.6.3 Conclusions

The staff review at the time of the initial licensing of a facility determined that the information provided by the applicant was sufficient for the staff to conclude that the evaluations of the effects of postulated pipe breaks both inside and outside containment met the staff acceptance criteria and the intent of all applicable regulations. The staff has and will continue to obtain new information related to this subject area through a variety of sources such as updates to the FSAR, research reports, operating plant events and routine plant inspections. The staff reviews this information and, in the past, has required licensees to take actions to upgrade the plant to provide continuing assurance of adequate protection of the public health and safety. In addition, the staff will continue to review new information in this subject area and if the staff determines that new or different requirements are needed, the staff has the capability within the existing regulatory process to require additional analyses or plant modifications, as necessary, to ensure the continued health and safety of the public. In conclusion, the Commission concludes that the current regulatory process has and will continue to provide reasonable assurance that the licensing basis of all currently operating plants are sufficient to assure that operation is not inimical to the public health and safety.

3.7 Seismic Design

3.7.1 Safety Issues and Regulatory Requirements

The Commission regulations require, in part, that systems, structures, and components important to safety be designed to withstand the effects of earthquakes without loss of capability to perform their safety functions and provides, in part, criteria required to determine the suitability of the plant design bases that were established by consideration of the seismic characteristics of the proposed plant site.

3.7.2 Evolution of Current Licensing Basis

The Commission has recognized the potential safety-related consequences of the occurrence of a significant seismic event at a nuclear power plant site since the staff first began reviewing applications for licenses in the late 1950s. These early reviews were performed on a plant-specific basis, and the staff's guidelines for acceptability evolved as new facilities were reviewed. 10 CFR Part 100, Appendix A, was later issued to establish the seismic design basis for systems, structures, and components.

In 1973, Regulatory Guides 1.60 and 1.61 were issued to provide staff positions relative to seismic input levels to be used for plant designs to ensure adequate consideration of historical data, site characteristics, and material behavior. In 1975, SRPs 3.7.1, 3.7.2, and 3.7.3 were issued to provide detailed guidelines for analytical modeling techniques for seismic analyses. These guidelines were used by the staff to determine that a plant-specific design satisfies applicable portions of the Commission's regulations. These SRPs were revised in 1981 to reflect changes in these guidelines since 1975. For example, one of the significant changes was related to the staff position on soil-structure interaction which was based on 1975 state-of-the-art. This position was widely debated among industry, ACRS, and the staff. These debates provided the basis for significant industry research effort on this issue during the 1970s. The staffs' evaluation of this research provided the basis for a change in the staff position which incorporated the research recommendations and led to more realistic criteria.

In the course of evaluating several plant-specific piping designs in the late 1970s, the staff became aware of significant discrepancies between the original piping seismic analysis computer code and a staff-approved benchmark code. This problem led to a March 13, 1979 Order to Show Cause from the Commission, which resulted in the shutdown of five plants whose piping designs had involved the use of the suspect computer codes. The differences between the computer codes were attributed to the use of an inappropriate method of combining certain seismic-induced loads in the original codes. In April 1979, IE Bulletin 79-07 was issued to request all licensees and applicants to review their piping analyses and determine if any of their computer codes contained the unacceptable method of combining loads. In addition, they were requested to verify that all piping computer programs were checked against either staff-approved benchmark problems or other acceptable piping computer programs. All licensees were to submit reports to the Commission describing the results of their review. The staff reviewed the submittals from all licensees and applicants and the issue was resolved on a plant-specific basis by arriving at one of the following conclusions:

1. The licensee or applicant used acceptable methods of combining loads in their piping analyses.

2. If the original analyses used the unacceptable method of combining loads, all applicable piping was reanalyzed using acceptable methodology. The results of these new analyses showed that all piping stresses were within the allowable stresses of applicable ASME Section III or ANSI B31.1 Codes.

3.7.3 Conclusions

The staff review at the time of the initial licensing of a facility determined that the information provided by the applicant was sufficient for the staff to conclude that the design for protection against earthquakes met the staff acceptance criteria and the intent of all applicable regulations. The staff has and will continue to obtain new information related to this subject area through a variety of sources such as updates to the FSAR, research reports, operating plant events and routine plant inspections. The staff reviews this information and, in the past, has required licensees to take actions to upgrade the plant to provide continuing assurance of adequate protection of the public health and safety. In addition, the staff will continue to review new information in this subject area and if the staff determines that new or different requirements are needed, the staff has the capability within the existing regulatory process to require additional analyses or plant modifications, as necessary, to ensure the continued health and safety of the public. In conclusion, the Commission concludes that the current regulatory process has and will continue to provide reasonable assurance that the licensing basis of all currently operating plants are sufficient to assure that operation is not inimical to the public health and safety.

3.8   Design of Seismic Category I Structures

3.8.1   Safety Issues and Regulatory Requirements

The Commission regulations contain various requirements for the design and construction of concrete and steel containments as well as requirements for all other Seismic Category I structures both inside and outside containment.

3.8.2   Evolution of Current Licensing Basis

When the NRC staff first began to review design of Seismic Category I structures in the 1960s, its scope and depth of review were not well defined and the staff acceptance was generally based on an ad hoc and plant-specific approach that provided adequate protection of the general public.

Staff review of design adequacy of containment and other Category I structures has been generally upgraded to use established structural design criteria and methodologies. The primary codes used in the early 1970s to review the design adequacy were the Building Code Requirements for Reinforced Concrete (ACI-318) and AISC, "Specification for Design, Fabrication, and Erection of Structural Steel for Buildings," American Institute of Steel Construction, for concrete and steel structures, respectively.

In 1973, the ASME Boiler and Pressure Vessel Code, Section III, Divisions 1 and 2, became the standards for the design of steel and concrete containments. In 1975, the staff published the Standard Review Plans (NUREG-0800), which adopted the above-listed codes and standards with appropriate inclusion of new load combinations and analysis methods. The Standard Review Plans were revised in 1981 and Sections 3.8.1 through 3.8.5 of the plans form the bulk of the current licensing criteria for containments and Category I structures.

The staff, as part of its routine review process, continued to assess the potential safety impact of any new information related to design of structures and, as appropriate, caused plant modifications to be implemented for affected plants to ensure continued conformance with the current licensing basis (e.g., modifications of torus supports and header piping supports for all Mark I plants).

3.8.3 Conclusions

The staff review at the time of the initial licensing of a facility determined that the information provided by the applicant was sufficient for the staff to conclude that the design of seismic catetory I structures met the staff acceptance criteria and the intent of all applicable regulations. The staff has and will continue to obtain new information related to this subject area through a variety of sources such as updates to the FSAR, research reports, operating plant events and routine plant inspections. The staff reviews this information and, in the past, has required licensees to take actions to upgrade the plant to provide continuing assurance of adequate protection of the public health and safety. In addition, the staff will continue to review new information in this subject area and if the staff determines that new or different requirements are needed, the staff has the capability within the existing regulatory process to require additional analyses or plant modifications, as necessary, to ensure the continued health and safety of the public. In conclusion, the Commission concludes that the current regulatory process has and will continue to provide reasonable assurance that the licensing basis of all currently operating plants are sufficient to assure that operation is not inimical to the public health and safety.

3.9 Mechanical Systems and Components

3.9.1 Safety Issues and Regulatory Requirements

The Commission regulations contain requirements to ensure that all of the different types of mechanical systems, components, and equipment will maintain their structural and functional integrity for the life of the plant.

3.9.2 Evolution of Current Licensing Basis

During the early 1960s when 10 CFR 50 General Design Criteria and the ASME Section III Code for Nuclear Power Plant Components were evolving, the staff reviews of design criteria for mechanical systems and components were performed on a plant-specific basis. ASME Section VIII, "Pressure Vessels," and American National Standards Institute (ANSI) B31.1, "Power Piping," were the two main design standards that were accepted by the staff to ensure the structural integrity of safety-related mechanical systems and components. In 1963, ASME Section III, "Nuclear Vessels," was published and accepted by the staff as a replacement for ASME Section VIII. In 1969, ANSI B31.7, "Nuclear Power Piping," was published and accepted by the staff as a replacement for ANSI B31.1. In 1971, ASME Section III was expanded to include rules for vessels, pumps, valves, and piping. ANSI B31.7 was included in ASME Section III-1971. In that same year, 10 CFR 50.55a was added to the regulations to provide a requirement for applicants to use ASME Section III for the design of reactor coolant pressure boundary components. Subsequent editions of ASME Section III through the present 1989 edition have been required by 10 CFR 50.55a for the designs of mechanical systems and components.

In 1975, Standard Review Plans 3.9.1 through 3.9.5 were issued to document guidelines that the staff had previously been using in its plant reviews of mechanical systems and components to demonstrate compliance with the applicable Commission regulations. SRP 3.9.6 was also issued in 1975 to provide guidelines for the staff to use in evaluating inservice testing (IST) programs for safety-related pumps and valves in non-operating plants. At that time, there was no requirement for IST applicable to operating plants. Therefore, in February 1976, 10 CFR 50.55a(g) was revised to include specific IST requirements in accordance with ASME Section XI for all licensees of operating plants and all applicants for a license to operate. Included in this revision was a requirement for all licensees to submit a new IST program to the staff every 10 years. These new programs are updated to reflect the latest ASME Section XI requirements and staff positions. Each licensee's program is reviewed and approved by the staff.

Prior to 1979, light water reactors experienced a number of occurrences of improper performance of safety and relief valves installed in the reactor coolant system. As a result in 1980, the staff issued NUREG-0737, "Clarification of TMI Action Plan Requirements," Item II.D.1, which required all BWR and PWR licensees and applicants to qualify reactor coolant system safety and relief valves, block valves, and associated piping and supports under expected operating conditions for design basis transients and accidents. In response to this requirement, the Electric Power Research Institute (EPRI) conducted a series of tests for licensees and applicants in 1981 and 1982 to demonstrate operability of the components under the required loading conditions. All licensees and applicants have submitted information to demonstrate applicability of the EPRI test results to their plant-specific equipment. The staff has reviewed all but a few plant-specific responses on this issue. The staff's reviews assure that all applicable valves, piping, and supports in each plant are enveloped by the EPRI test program. These reviews are scheduled for completion in 1990.

3.9.3 Conclusions

The staff review at the time of the initial licensing of a facility determined that the information provided by the applicant was sufficient for the staff to conclude that the design of mechanical systems and components important to safety met the staff acceptance criteria and the intent of all applicable regulations. The staff has and will continue to obtain new information related to this subject area through a variety of sources such as updates to the FSAR, research reports, operating plant events and routine plant inspections. The staff reviews this information and, in the past, has required licensees to take actions to upgrade the plant to provide continuing assurance of adequate protection of the public health and safety. In addition, the staff will continue to review new information in this subject area and if the staff determines that new or different requirements are needed, the staff has the capability within the existing regulatory process to require additional analyses or plant modifications, as necessary, to ensure the continued health and safety of the public. In conclusion, the Commission concludes that the current regulatory process has and will continue to provide reasonable assurance that the licensing basis of all currently operating plants are sufficient to assure that operation is not inimical to the public health and safety.

3.10 Seismic and Dynamic Qualification of Mechanical and Electrical Equipment

3.10.1 Safety Issues and Regulatory Requirements

The Commission regulations contain requirements to ensure that mechanical and electrical equipment important to safety remain operable under the full range of normal and accident loadings, including seismic.

3.10.2 Evolution of Current Licensing Basis

The evolution of the GDC, safety guides, regulatory guides, and SRPs is discussed in Subsection 3.2.2 above. Commission guidance was originally issued in March 1976. These guidelines documented staff positions on seismic and dynamic qualification of equipment that had been implemented by the staff since the early 1970s. The analysis and test criteria used by the staff to review this issue evolved rapidly between 1971 and 1980. Consequently, for plants that were reviewed by the staff prior to the early 1970s, the margins of safety provided in equipment to resist seismically induced loads are uncertain. This concern led the staff to originate Unresolved Safety Issue (USI) A-46, "Seismic Qualification of Equipment in Operating Nuclear Power Plants."

NUREG-1211, "Regulatory Analysis for Resolution of USI A-46," February 1987, identified some operating plants to be reviewed under USI A-46 as those plants whose equipment had not been qualified by using IEEE Standard 344-1975 or later revision. On February 19, 1987, the staff issued Generic Letter 87-02, "Verification of Seismic Adequacy of Mechanical and Electrical Equipment in Operating Reactors, USI A-46." This letter provided the staff's requirements for implementing the resolution of USI A-46. A utility owners group has been formed and is negotiating with the staff on the implementation schedule in accordance with the NRC policy on integrated schedule for plant modifications, Generic Letter 83-20, May 9, 1983. In addition, the staff will conduct detailed audits of a limited number of plants to verify that the licensee has implemented its program in accordance with GL 87-02. Enforcement or other regulatory actions could result from these audits. When a licensee completes its followup actions, it will be required to submit a letter stating that all plant modifications or followup actions related to USI A-46 have been completed.

3.10.3 Conclusions

The staff review at the time of the initial licensing of a facility determined that the information provided by the applicant was sufficient for the staff to conclude that the seismic and dynamic qualification of mechanical and electrical equipment met the staff acceptance criteria and the intent of all applicable regulations. The staff has and will continue to obtain new information related to this subject area through a variety of sources such as updates to the FSAR, research reports, operating plant events and routine plant inspections. The staff reviews this information and, in the past, has required licensees to take actions to upgrade the plant to provide continuing assurance of adequate protection of the public health and safety. In addition, the staff will continue to review new information in this subject area and if the staff determines that new or different requirements are needed, the staff has the capability within the existing regulatory process to require additional analyses or plant modifications, as necessary, to ensure the continued health and safety of the public. In conclusion, the Commission concludes that the current regulatory process has and will continue to provide reasonable assurance that the licensing basis of all currently operating plants are sufficient to assure that operation is not inimical to the public health and safety.

3.11 Environmental Design of Mechanical and Electrical Equipment

3.11.1 Safety Issues and Regulatory Requirements

The Commission regulations require that systems, structures, and components important to safety be designed to accommodate the effects of and be compatible with the environmental conditions associated with normal operation, maintenance, testing, and postulated accidents, including loss-of-coolant accidents. They are to be appropriately protected against the effects of discharging fluids. Environmental qualification is one means of satisfying the above requirement for essential components. Specific requirements for environmental qualification of electrical equipment important to safety are contained in 10 CFR 50.49, which requires each licensee to establish a program for qualification of essential electrical equipment subject to harsh environmental conditions, and maintain qualification of this equipment for the lifetime of the plant.

For purposes of the discussion on the environmental design basis, it should be noted that licensees' current environmental qualification programs will include equipment subject to periodic replacement and equipment that has been qualified for the currently licensed plant lifetime. A review of the program covering equipment periodically replaced is not necessary as this process will continue during the renewed license life.

3.11.2 Evolution of Current Licensing Basis

In November 1977, the Union of Concerned Scientists petitioned the Commission to upgrade the environmental qualification of equipment in operating facilities to current standards. This petition led to the Commission Memorandum and Order of May 23, 1980 (CLI-80-21) which provided guidance and directives to resolve this matter in an expeditious manner. Part of this activity included development of a new rule for environmental qualification of electrical equipment. This action culminated in issuance of 10 CFR 50.49 dated January 21, 1983. All licensees have implemented programs consistent with 10 CFR 50.49 and supplemental staff guidelines to ensure the safety function of electrical equipment subjected to harsh environments (radiation, temperature, pressure, and moisture) following postulated design basis accidents. This has resulted in assurance of safe plant shutdown following loss of coolant and steam line break accidents.

3.11.3 Conclusions

The staff review at the time of the initial licensing of a facility determined that the information provided by the applicant was sufficient for the staff to conclude that the environmental qualification program for electrical equipment important to safety met the staff acceptance criteria and the intent of all applicable regulations. The staff has and will continue to obtain new information related to this subject area through a variety of sources such as updates to the FSAR, research reports, operating plant events and routine plant inspections. The staff reviews this information and, in the past, has required licensees to take actions to upgrade the plant to provide continuing assurance of adequate protection of the public health and safety. In addition, the staff will continue to review new information in this subject area and if the staff determines that new or different requirements are needed, the staff has the capability within the existing regulatory process to require additional analyses or plant modifications, as necessary, to ensure the continued health and safety of the public. In conclusion, the Commission concludes that the current regulatory process has and will continue to provide reasonable assurance that the licensing basis of all currently operating plants are sufficient to assure that operation is not inimical to the public health and safety.

4 REACTOR

4.1 Scope

This chapter addresses the evaluation and supporting information reviewed by the staff to establish the capability of the reactor to perform its safety functions throughout its design lifetime under all normal operational modes, including transient and steady state, and accident conditions. The evaluation covers the areas of fuel system design, nuclear design, thermal and hydraulic design, reactor materials, and functional design of reactivity control systems.

4.2 Fuel System Design

4.2.1 Safety Issues and Regulatory Requirements

Commission regulations require that the reactor core and associated coolant, control, and protection systems be designed with appropriate margin to ensure that specified acceptable fuel design limits (SAFDLs) are not exceeded during any condition of normal operation, including the effects of anticipated operational occurrences, and that the reactor core and associated coolant systems be designed so that in the power operating range the net effect of the prompt inherent nuclear feedback characteristics tends to compensate for a rapid increase in reactivity. These regulations also address the requirements of maintaining the capability to cool the core under postulated accident conditions. Methods of adequately predicting fuel rod failures during postulated accidents are adopted so that radioactivity release estimates are not underestimated and thereby ensure that the plant in question would continue to satisfy the related requirements of 10 CFR Part 100. Also, the acceptable fuel performance limits under a postulated loss of coolant accident are specified in 10 CFR 50.46 and Appendix K to 10 CFR Part 50.

4.2.2 Evolution of Current Licensing Basis

The evolution of the general design criteria (GDC), safety guides, regulatory guides, and standard review plans (SRPs) is discussed in Section 3.2.2 above.

4.2.3 Conclusions

The staff review at the time of the initial licensing and before each refueling of a facility determined that the information provided by the applicant was sufficient for the staff to conclude that the fuel system design met the staff acceptance criteria and the intent of all applicable regulations. The staff has and will continue to obtain new information related to this subject area through a variety of sources such as updates to the FSAR, research reports, operating plant events and routine plant inspections. The staff reviews this information and, in the past, has required licensees to take actions to upgrade the plant to provide continuing assurance of adequate protection of the public health and safety. In addition, the staff will continue to review new information in this subject area and if the staff determines that new or different requirements are needed, the staff has the capability within the existing regulatory process to require additional analyses or plant modifications, as necessary, to ensure the continued health and safety of the public. In conclusion, the Commission concludes that the current regulatory process has and will continue to provide reasonable assurance that the licensing basis of all currently operating plants are sufficient to assure that operation is not inimical to the public health and safety.

4.3 Nuclear Design

4.3.1 Safety Issues and Regulatory Requirements

Commission regulations require that acceptable fuel design limits be specified that will not be exceeded during normal operation, including the effects of anticipated operational occurrences, and require that, in the power operating range, the prompt inherent nuclear feedback characteristics tend to compensate for a rapid increase in reactivity. The regulations also require that power oscillations that could result in conditions exceeding SAFDL are not possible or can be reliably and readily detected and suppressed; and require instrumentation and controls to monitor variables and systems that can affect the fission process over anticipated ranges for normal operation and accident conditions, and to maintain the variables and systems within prescribed operating ranges. Further, the regulations require automatic initiation of the reactivity control systems to ensure that SAFDLs are not exceeded; that reliable reactivity control systems be provided under normal or accident operating conditions; and that the effects of postulated reactivity accidents neither result in damage to the reactor coolant pressure boundary greater than limited local yielding, nor cause sufficient damage to impair significantly the capability to cool the core.

4.3.2 Evolution of Current Licensing Basis

The evolution of the GDC, safety guides, regulatory guides, and SRPs is discussed in Section 3.2.2 above.

4.3.3 Conclusions

The staff review at the time of the initial licensing and before each refueling of a facility determined that the information provided by the applicant was sufficient for the staff to conclude that the nuclear design met the staff acceptance criteria and the intent of all applicable regulations. The staff has and will continue to obtain new information related to this subject area through a variety of sources such as updates to the FSAR, research reports, operating plant events and routine plant inspections. The staff reviews this information and, in the past, has required licensees to take actions to upgrade the plant to provide continuing assurance of adequate protection of the public health and safety. In addition, the staff will continue to review new information in this subject area and if the staff determines that new or different requirements are needed, the staff has the capability within the existing regulatory process to require additional analyses or plant modifications, as necessary, to ensure the continued health and safety of the public. In conclusion, the Commission concludes that the current regulatory process has and will continue to provide reasonable assurance that the licensing basis of all currently operating plants are sufficient to assure that operation is not inimical to the public health and safety.

4.4 Thermal and Hydraulic Design

4.4.1 Safety Issues and Regulatory Requirements

Commission regulations require that the reactor core and associated coolant, control, and protection systems be designed with appropriate margin to ensure that specified acceptable fuel design limits (SAFDLs) are not exceeded during any condition of normal operation, including the effects of anticipated operational occurrences.

4.4.2 Evolution of Current Licensing Basis

The evolution of the GDC, safety guides, regulatory guides, and SRPs is discussed in Section 3.2.2 above.

4.4.3 Conclusions

The staff review at the time of the initial licensing and before each refueling of a facility determined that the information provided by the applicant was sufficient for the staff to conclude that the thermal and hydraulic design met the staff acceptance criteria and the intent of all applicable regulations. The staff has and will continue to obtain new information related to this subject area through a variety of sources such as updates to the FSAR, research reports, operating plant events and routine plant inspections. The staff reviews this information and, in the past, has required licensees to take actions to upgrade the plant to provide continuing assurance of adequate protection of the public health and safety. In addition, the staff will continue to review new information in this subject area and if the staff determines that new or different requirements are needed, the staff has the capability within the existing regulatory process to require additional analyses or plant modifications, as necessary, to ensure the continued health and safety of the public. In conclusion, the Commission concludes that the current regulatory process has and will continue to provide reasonable assurance that the licensing basis of all currently operating plants are sufficient to assure that operation is not inimical to the public health and safety.

4.5 Reactor Materials

4.5.1 Safety Issues and Regulatory Requirements

Commission regulations, in part, require that the reactor coolant pressure boundary have an extremely low probability of abnormal leakage, of rapidly propagating failure, and of gross rupture under operating, maintenance, testing, and postulated accident conditions. These regulations also require that structures, systems, and components important to safety be designed, fabricated, erected, and tested to quality standards commensurate with the importance of the safety functions to be performed. In addition, the regulations require that one of the reactivity control systems shall use control rods, preferably including a positive means for inserting the rods, and shall be capable of reliably controlling reactivity changes to ensure that fuel design limits are not exceeded under conditions of normal operation, including anticipated operational occurrences, and that components that are part of the reactor coolant pressure boundary be designed to permit periodic inspection and testing of critical areas to assess their structural and leaktight integrity.

To satisfy these regulations, the staff recommends that control rod drive structural materials, reactor internals, and core support materials be designed, fabricated, erected, and tested using the published regulatory guidance and inspected to the guidelines of Section XI, Code Class 1, of the American Society of Mechanical Engineers Boiler and Pressure Vessel Code (hereafter ASME Code). The staff reviews proposed alternatives to the recommendations in the current criteria to ensure that they provide an acceptable level of quality and safety.

When inservice inspection requirements of Section XI of the ASME Code are determined to be impractical, the NRC, in accordance with 10 CFR 50.55a(g)(6), may grant relief and may impose alternative requirements that are determined to be authorized by law, will not endanger life or property or the common defense and security, and are otherwise in the public interest, giving due consideration to the burden upon the licensee that could result if the requirements were imposed on the facility.

The staff has implemented requirements in addition to Section III of the ASME Code because many of the components are constructed of austenitic stainless steel material that is susceptible to intergranular stress corrosion cracking in the water environment of the boiling-water reactor (BWR).

4.5.2 Evolution of Current Licensing Basis

Sections III and XI of the ASME Code have changed and will continue to change as technology and operating experience change. The staff actively participates in the process that revises the ASME Code and reviews these changes to determine whether they should be incorporated into plants' licensing bases.

Section XI of the ASME Code contains updating provisions. It requires licensees to revise their inservice inspection program every 10 years. The revised programs incorporate all the changes required by ASME Section XI of the licensee's program and, in accordance with 10 CFR 50.55a(g)(6), the staff may grant relief or impose alternative requirements.

Changes in technologies (i.e., radiation embrittlement, ultrasonic examination, etc.), which are not addressed by the ASME Code, are described in generic letters and regulatory guides. These letters and guides are prepared and published by the staff and become incorporated into the licensing basis of any nuclear power plant. Many of these guides recommend changes to the plant-specific licensing basis when environmental conditions change. The staff reviews these licensing changes and approves them on a plant-specific basis using generic acceptance criteria.

4.5.3 Conclusions

The staff review at the time of the initial licensing of a facility determined that the information provided by the applicant was sufficient for the staff to conclude that the control rod drive structural materials and reactor internals met the staff acceptance criteria and the intent of all applicable regulations. The staff has and will continue to obtain new information related to this subject area through a variety of sources such as updates to the FSAR, research reports, operating plant events and routine plant inspections. The staff reviews this information and, in the past, has required licensees to take actions to upgrade the plant to provide continuing assurance of adequate protection of the public health and safety. In addition, the staff will continue to review new information in this subject area and if the staff determines that new or different requirements are needed, the staff has the capability within the existing regulatory process to require additional analyses or plant modifications, as necessary, to ensure the continued health and safety of the public. In conclusion, the Commission concludes that the current regulatory process has and will continue to provide reasonable assurance that the licensing basis of all currently operating plants are sufficient to assure that operation is not inimical to the public health and safety.

4.6 Functional Design of Reactivity Control Systems

4.6.1 Safety Issues and Regulatory Requirements

Commission regulations require that the protection system be designed to fail into a safe state or into a state demonstrated to be acceptable on some other defined basis if such conditions as disconnection of the system, loss of energy (e.g., electric power, instrument air), or postulated adverse environments (e.g., extreme heat or cold, fire, pressure, steam, water, radiation) are experienced. These regulations also require that the protection system be designed to ensure that SAFDLs are not exceeded for any single malfunction of reactivity control systems, such as accidental withdrawal of control rods, and require that two independent reactivity control systems of different design principles be provided.

One of the systems uses control rods, preferably including a positive means for inserting the rods, and shall be capable of reliably controlling reactivity changes to ensure that under conditions of normal operation, including anticipated operational occurrences, and with appropriate margin for such malfunctions as stuck rods, SAFDLs are not exceeded. The second reactivity control system shall be capable of reliably controlling the rate of reactivity changes resulting from planned, normal power changes (including xenon burnout) to ensure that SAFDLs are not exceeded. One of the systems shall be capable of holding the reactor core subcritical under cold conditions.

Further, these regulations require that the reactivity control systems be designed to have a combined capability, in conjunction with poison addition by the emergency core cooling system, of reliably controlling reactivity changes to ensure that under postulated accident conditions and with appropriate margin for stuck rods the capability to cool the core is maintained; that the reactivity control systems be designed to consider the effects of postulated reactivity accidents and maintain core coolability; and that the reactivity control systems be designed to ensure an extremely high probability of accomplishing their safety function in the event of anticipated operational occurrences.

4.6.2 Evolution of Current Licensing Basis

The evolution of the GDC, safety guides, regulatory guides, and SRPs is discussed in Section 3.2.2 above.

4.6.3 Conclusions

The staff review at the time of the initial licensing and before refueling of a facility determined that the information provided by the applicant was sufficient for the staff to conclude that the functionsl design of reactivity control systems met the staff acceptance criteria and the intent of all applicable regulations. The staff has and will continue to obtain new information related to this subject area through a variety of sources such as updates to the FSAR, research reports, operating plant events and routine plant inspections. The staff reviews this information and, in the past, has required licensees to take actions to upgrade the plant to provide continuing assurance of adequate protection of the public health and safety. In addition, the staff will continue to review new information in this subject area and if the staff determines that new or different requirements are needed, the staff has the capability within the existing regulatory process to require additional analyses or plant modifications, as necessary, to ensure the continued health and safety of the public. In conclusion, the Commission concludes that the current regulatory process has and will continue to provide reasonable assurance that the licensing basis of all currently operating plants are sufficient to assure that operation is not inimical to the public health and safety.

5 REACTOR COOLANT SYSTEM AND CONNECTED SYSTEMS

5.1 <u>Integrity of Reactor Coolant Pressure Boundary and Reactor Vessels</u>

5.1.1 Scope

This section addresses an evaluation of the reactor coolant system and systems connected to it. Special consideration is given to the reactor coolant system and pressure-containing parts out to and including isolation valving which is the reactor coolant pressure boundary (RCPB), as defined in 10 CFR 50.2(v). The evaluation covers the areas of integrity of RCPB, reactor vessels, and component and subsystem design.

5.1.2 Safety Issues and Regulatory Requirements

Commission regulations require that the RCPB have an extremely low probability of abnormal leakage, of rapidly propagating failure, and of gross rupture under operating maintenance, testing, and postulated accident conditions and require that the reactor coolant system and associated auxiliary, control, and protection systems be designed with sufficient margin to ensure that the design conditions of the RCPB are not exceeded during any conditions of normal operation, including anticipated operational occurrences. These regulations also require that components that are part of the RCPB be designed, fabricated, erected, and tested to the highest quality standards practical and also that they be designed to permit periodic inspection and testing of critical areas to assess their structural and leaktight integrity.

To satisfy these requirements, RCPB components must be designed, fabricated, erected, and tested to 10 CFR 50.55a(c), inservice inspection must be performed according to 10 CFR 50.55a(g), and RCPB components must meet the fracture toughness and material surveillance requirements of 10 CFR 50.60. Additional fracture toughness requirements for protection against pressurized thermal shock events are contained in 10 CFR 50.61. Also protection against overpressure is provided per the requirements of American Society of Mechanical Engineers Boiler and Pressure Vessel Code (ASME Code), Section III, Article NB-7000.

10 CFR 50.55a(c) requires RCPB components to meet the edition and addenda of Section III of the ASME Code that was required by Commission regulations at the time the regulations were issued. 10 CFR 50.55a(g) requires RCPB components to meet Section XI of the ASME Code. Proposed alternatives to these ASME Code requirements are permitted in 10 CFR 50.55(a)(3), provided the licensee demonstrates that (1) the proposed alternative would provide an acceptable level of quality and safety or (2) compliance with the specified requirements would result in hardship or unusual difficulties without a compensating increase in the level of quality and safety. When inservice inspection requirements of Section XI of the ASME Code are determined to be impractical, the NRC, in accordance with 10 CFR 50.55a(g)(6), may grant relief and may impose alternative requirements that are determined to be authorized by law, will not endanger life or property or the common defense and security, and are otherwise in the public interest, giving due consideration to the burden upon the licensee that could result if the requirements were imposed on the facility.

10 CFR 50.60(a) requires that RCPB components meet the fracture toughness requirements in Appendix G to 10 CFR Part 50, and requires the reactor vessel material surveillance program to meet the requirements in Appendix H to 10 CFR Part 50. These appendices impose additional requirements on the reactor vessel because the reactor vessel is subject to neutron irradiation embrittlement. 10 CFR 50.60(b) permits licensees to meet alternative requirements to those specified in Appendices G and H when an exemption is granted by the Commission under 10 CFR 50.12. A low-temperature overpressure protection (LTOP) system is provided to ensure that the pressure-temperature limits per the Appendix G requirements are not exceeded.

5.1.3 Evolution of Current Licensing Basis

The RCPB is composed of piping, pumps, valves, and vessels. Before 1970, piping in the RCPB was constructed and fabricated to the American National Standards Institute (ANSI) Code B31.1, "Power Piping," and pumps and valves were constructed and fabricated to manufacturer specifications. In 1970, the ASME Code was revised to include requirements for RCPB piping, pumps, and valves in Section III. Vessels within the RCPB are constructed and fabricated to ASME Code requirements. Earlier plants were constructed to Sections I and VIII, and later plants were constructed to Section III requirements.

The fabrication requirements for RCPB piping, pumps, valves, and vessels are specified in the plant's final safety analysis report (FSAR). The staff reviewed the plant's FSAR to determine that the alternative requirements to Section III of the ASME Code provided an acceptable level of quality and safety.

Sections III and XI of the ASME Code have changed and will keep changing during the licensed lifetime of nuclear power plants including the renewal period. The staff reviews these changes to determine whether they should be incorporated into the individual plant licensing bases.

The most significant change in Section III of the ASME Code, which affects RCPB integrity, was a change in fracture toughness requirements initiated in the Summer 1972 Addenda to the 1971 Edition of the ASME Code. This change required additional material testing and required pressure/temperature (P/T) limits during heatup, cooldown, and hydrotest of the reactor vessel. All licensees are required to heat up, cool down, and hydrotest the reactor vessel in accordance with plant-specific P/T limits that are based on linear elastic fracture mechanics technology. However, the Commission decided that plants built before 1972 and plants that had ordered their reactor coolant pressure boundary material before 1972 did not have to perform the additional testing required by the Summer 1972 Addenda. Except for reactor vessel materials, the Commission concluded that the earlier test requirements were adequate to ensure RCPB integrity. Commission acceptance of these requirements is documented in Appendix G to 10 CFR Part 50. For reactor vessel materials, the staff issued Branch Technical Position (MTEB) 5-2, "Fracture Toughness Requirement." This branch technical position described a method of updating the earlier test data to Summer 1972 Addenda requirements. The test data must be updated in order to calculate P/T limits. Licensees utilized this method or developed their own method to update their reactor vessel material test data. Methods developed by licensees were reviewed and approved by the staff.

10 CFR 50.55a(g) requires that RCPB components meet Section XI of the ASME Code. Section XI of the ASME Code contains updating provisions. It requires licensees to revise their inservice inspection programs every 10 years. The revised programs incorporate all the changes required by Section XI of the ASME Code, except for those that are impractical. The staff reviews the licensee's program and, in accordance with 10 CFR 50.55a(g)(6), the staff may grant relief or impose alternative requirements.

Changes in technologies (i.e., radiation embrittlement, ultrasonic examination, etc.), which are not addressed by the ASME Code, are described in generic letters and regulatory guides. These letters and guides are prepared by the staff and may be incorporated into the licensing basis of the nuclear power plant. Many of these guides recommend changes to the plant licensing basis when environmental conditions change. These licensing changes are reviewed and approved by the staff.

In addition to these requirements. the staff may recommend or require additional programs when it determines that the operating environment for the component is particularly severe. These programs are imposed through issuance of technical specifications or are recommended through issuance of branch technical positions, regulatory guides, standard review plans, or generic letters. Examples of components that operate in a particularly severe environment and for which the staff has either recommended or imposed additional requirements are boiling-water reactor (BWR) coolant pressure boundary piping, pressurized-water reactor (PWR) steam generator tubing, and all light-water reactor (LWR) vessels. Generic Letter 88-01 specified additional recommendations for BWR coolant pressure boundary piping because the piping was subject to intergranular stress corrosion cracking (IGSCC). Generic Letter 88-11 recommended a revised method of calculating neutron irradiation embrittlement of LWR reactor vessels because analysis of surveillance data in Appendix H to 10 CFR Part 50 indicated that the previous method did not adequately address the issue. The staff imposes augmented inspection program requirements on PWR steam generator tubing by issuing technical specifications.

The staff reviewed the capability to remove contaminants from the reactor coolant system. The reactor water cleanup systems in direct-cycle BWR plants, in conjunction with the primary water monitoring system, must have the capability to remove contaminants introduced by main condenser leakage. Since the SEP effort, the staff has encouraged the preparation and implementation of Electric Power Research Institute (EPRI) and BWR owners group normal and hydrogen water chemistry guidelines. These guidelines, implemented by all BWR owners, specifically address action recommendations in the event of condenser in-leakage transient implementation of the water chemistry guidelines through the use of NRC Inspection Procedure 79501. On the basis of such regulatory activities, the staff has concluded that the SEP lesson-learned issue has been acceptably resolved and that no additional regulatory action is necessary to address this issue.

A reliable and sensitive leakage detection system is important to monitor the reactor coolant pressure boundary leakage to the containment and interconnecting system and it provides operators with an adequate margin of time to initiate actions to identify, isolate, and repair the source of a leak. Revisions to procedures or technical specifications were made for some plants which reflected system limitations and to enhance system reliability. Generic Letters 84-04 and

88-01 addressed improved technical specifications for leakage detection on PWRs and BWRs, respectively. All plants have technical specifications that address allowable rates for leaks in the primary coolant system. Additionally, procedures and operator training improvements also address the detection of primary coolant system leakage and necessary actions.

5.1.4 Conclusions

The staff review at the time of the initial licensing of a facility determined that the information provided by the applicant was sufficient for the staff to conclude that the reactor coolant pressure boundary as defined in 10 CFR 50.2(n) met the staff acceptance criteria and the intent of all applicable regulations. The staff has and will continue to obtain new information related to this subject area through a variety of sources such as updates to the FSAR, research reports, operating plant events and routine plant inspections. The staff reviews this information and, in the past, has required licensees to take actions to upgrade the plant to provide continuing assurance of adequate protection of the public health and safety. In addition, the staff will continue to review new information in this subject area and if the staff determines that new or different requirements are needed, the staff has the capability within the existing regulatory process to require additional analyses or plant modifications, as necessary, to ensure the continued health and safety of the public. In conclusion, the Commission concludes that the current regulatory process has and will continue to provide reasonable assurance that the licensing basis of all currently operating plants are sufficient to assure that operation is not inimical to the public health and safety.

5.2 Component and Subsystem Design

5.2.1 Scope

This section addresses the performance requirements and design features to ensure overall safety of the various components within the reactor coolant system and subsystems closely allied with the reactor coolant system. This component and subsystem include reactor coolant pumps, steam generators, reactor coolant piping, main steamline flow restrictions, main steamline isolation system, reactor core isolation cooling system, residual heat removal system, reactor water cleanup system, main steamline and feedwater piping, pressurizer, pressurizer relief discharge system, valves, safety and relief valves, component supports, and reactor coolant system high point vents.

5.2.2 Safety Issues and Regulatory Requirements

Commission regulations require that systems, structures, and components important to safety be designed, fabricated, erected, and tested to quality standards commensurate with the importance of the safety functions to be performed and that systems, structures, and components important to safety shall be appropriately protected against dynamic effects, including the effects of missiles, pipe whipping, and discharging fluids, that may result from equipment failures and from events and conditions outside the plant. In addition, the regulations require (1) that the RCPB shall be designed, fabricated, erected, and tested so as to have an extremely low probability of abnormal leakage, of rapidly propagating failure, and of gross rupture; (2) that the reactor coolant system and associated auxiliary systems shall be designed with sufficient margin to ensure that the design conditions of the RCPB are not exceeded during any condition of normal operation, including anticipated operational occurrences; (3) that the RCPB shall be designed with sufficient margin to ensure that, when stressed under operating, maintenance, testing, and postulated accident conditions, the boundary behaves in a nonbrittle manner and the probability of rapidly propagating fracture is minimized; (4) that components that are part of the RCPB shall be designed to permit periodic inspection and testing of important areas and features to assess their structural and leaktight integrity; and (5) that a system to supply reactor coolant makeup for protection against small breaks in the RCPB shall be provided.

The system safety function shall be to transfer fission product decay heat and other residual heat from the reactor core at such a rate that specified acceptable fuel design limits (SAFDLs) and the design conditions of the RCPB are not exceeded. Suitable redundancy in components and features and suitable interconnections, leak detection, and isolation capabilities shall be provided.

5.2.3 Evolution of Current Licensing Basis

The licensing basis for various components within the reactor coolant system and subsystems has evolved as reactor events and generic studies by the NRC staff provide new information that is determined to improve component and subsystem performance. The process of evaluating operating experience and assessing plant data to determine the need for additional actions is a continuous one.

From operating experiences, many different forms of steam generator tube degradation have been identified, including stress corrosion cracking, wastage, intergranular attack, denting, erosion-corrosion, fatigue cracking, pitting, fretting, support plate degradation, and mechanical damage resulting from impingement of foreign objects or loose parts on the internal components of steam generators. These degradations have resulted in extensive steam generator inspections, tube plugging, repair, or replacement. Also steam generator tube rupture (SGTR) events have occurred in a few operating reactors. Steam generator tube integrity was designated an unresolved safety issue (USI) in 1978 and Task Action Plans (TAPs) A-3, A-4, and A-5 were established to evaluate the safety significance of degradation in steam generators of various designs. NUREG-0844 was published in September 1988 to present the results of the NRC integrated program for the resolution of USIs A-3, A-4, and A-5 regarding steam generator tube integrity. A generic risk assessment is provided and indicates that risk from SGTR events is not a significant contributor to total risk at a given site, nor to the total risk to which the general public is routinely exposed. This finding is considered to be indicative of the effectiveness of licensee programs and regulatory requirements for ensuring steam generator tube integrity in accordance with Appendices A and B to 10 CFR Part 50. This report also identifies a number of staff-recommended actions that can further improve the effectiveness of licensee programs in ensuring the integrity of steam generator tubes and in mitigating the consequences of an SGTR. As part of the integrated program, the staff issued Generic Letter 85-02 encouraging licensees of PWRs to upgrade their programs, as necessary, to meet the intent of the staff-recommended actions; however, such actions do not constitute NRC requirements. In addition, the staff is pursuing a number of actions and studies involving steam generator issues to gain added assurance that risk from SGTR events will continue to be small.

Following the accident at Three Mile Island, Unit 2 (TMI-2), the staff found that additional means were necessary to vent noncondensible gases from the reactor coolant system which may inhibit core cooling during natural circulation. On the basis of knowledge gained from the TMI-2 accident, Item II.B.1, "Reactor Coolant System Vents," was incorporated into the licensing bases for individual plants when all operating nuclear power plants were required to implement reactor coolant system high point venting capability in accordance with these guidelines.

Also from the experience of the TMI-2 accident, the staff found that operational performance of the relief and safety valves under various operating conditions is significant to safety. Performance testing of BWR and PWR relief and safety valves was incorporated in individual plant licensing bases when all nuclear power plants were required to implement testing requirements in accordance with the guidelines contained in TMI Action Plan Item II.D.1.

5.2.4 Conclusions

The staff review at the time of the initial licensing of a facility determined that the information provided by the applicant was sufficient for the staff to conclude that the components within the reactor coolant system and subsystems met the staff acceptance criteria and the intent of all applicable regulations. The staff has and will continue to obtain new information related to this subject area through a variety of sources such as updates to the FSAR, research reports, operating plant events and routine plant inspections. The staff reviews this information and, in the past, has required licensees to take actions to upgrade the plant to provide continuing assurance of adequate protection of the public health and safety. New criteria and improvements necessary for safety which have resulted from the continuous staff review have included steam generator tube intergrity (GL 85-02), reactor coolant system vents (NUREG-0737, Item II.B.1), and performance testing of relief and safety valves (NUREG-0737, Item II.D.1). In addition, the staff will continue to review new information in this subject area and if the staff determines that new or different requirements are needed, the staff has the capability within the existing regulatory process to require additional analyses or plant modifications, as necessary, to ensure the continued health and safety of the public. In conclusion, the Commission concludes that the current regulatory process has and will continue to provide reasonable assurance that the licensing basis of all currently operating plants are sufficient to assure that operation is not inimical to the public health and safety.

6 ENGINEERED SAFETY FEATURES

6.1 Scope

Engineered safety features (ESFs) are provided to mitigate the consequences of postulated accidents in spite of the fact that these accidents are very unlikely. The engineered safety features included in plant designs vary depending on the type of plant (PWR or BWR) under evaluation. This section will discuss five general categories of features routinely considered under the subject of ESFs. These include: metallic and organic materials, containment systems, emergency core cooling systems, habitability systems, and fission product removal and control systems.

6.2 Metallic and Organic Materials

6.2.1 Safety Issues and Regulatory Requirements

The Commission regulations require that the containment boundary be designed with sufficient margin to ensure that, under operating, maintenance, testing, and postulated accident conditions, its ferritic materials behave in a nonbrittle manner and the probability of a rapidly propagating fracture is minimized. In addition, the regulations require that systems, structures, and components important to safety be designed, fabricated, erected, and tested to quality standards commensurate with the importance of the safety function performed. Specific guidance on satisfying these requirements is contained in applicable regulatory guides that refer to the criteria of ASME Section III for metallic materials used in ESF system construction. In specific cases, with proper justification, the staff evaluated and found acceptable alternatives to these criteria that continue to ensure ESF system integrity and performance.

In addition, 10 CFR 50.55a(g) requires that essential components in ESF systems built to ASME Section III criteria receive regularly scheduled inservice inspection in accordance with the criteria of ASME Section XI. Relief can be granted against the criteria of ASME Section XI when the NRC staff determines that alternative measures are in place to ensure fracture prevention of the pressure boundary.

6.2.2 Evolution of Current Licensing Basis

Sections III and XI of the ASME Code have changed and will continue to change during the plant lifetime of nuclear power plants based on operating experience. The staff reviews these changes to determine whether they should be incorporated into the licensing basis of operating plants.

Section XI of the ASME Code contains updating provisions. It requires licensees to revise their inservice inspection program every 10 years. The revised programs incorporate all changes required by Section XI of the licensee's program and, in accordance with 10 CFR 50.55a(g)(6), the staff may grant relief or impose alternative requirements.

The staff reviewed the use of organic materials inside containment. The basis for the selection of paints and other organic materials is not documented for most operating reactors. The plant design must assure that organic materials, such as organic paints, coatings and insulation materials, used inside containment do not adversely affect the operation of the engineered safety feature equipment inside containment during accidents when they may be exposed to high temperatures, steam environments, high radiation fields, and containment spray systems. Since the completion of the SEP effort, the staff has implemented some actions related to this topic: Regulatory Guide 1.54, which ensorsed industry standards ANSI N101.4 (1972), "Quality Assurance for Protective Coatings Applied to Nuclear Facilities," and ANSI N101.2 (1927), "Quality Assurance Program Requirements for Nuclear Power Plants." In addition the NRC issued NUREG-0897, "Containment Emergency Sump Performance," and NUREG/CR-2791, "Methodology for Evaluation of Insulation Debris Effects," to provide a basis for making an evaluation of sump performance and insulation debris effects.

Based upon the above and industry actions, the staff concludes that adequate guidance and information has been issued and that licensee actions are adequate to address this issue.

6.2.3 Conclusions

The staff review at the time of the initial licensing of a facility determined that the information provided by the applicant was sufficient for the staff to conclude that ESF system components met the staff acceptance criteria and the intent of all applicable regulations. The staff has and will continue to obtain new information related to this subject area through a variety of sources such as updates to the FSAR, research reports, operating plant events and routine plant inspections. The staff reviews this information and, in the past, has required licensees to take actions to upgrade the plant to provide continuing assurance of adequate protection of the public health and safety. In addition, the staff will continue to review new information in this subject area and if the staff determines that new or different requirements are needed, the staff has the capability within the existing regulatory process to require additional analyses or plant modifications, as necessary, to ensure the continued health and safety of the public. In conclusion, the Commission concludes that the current regulatory process has and will continue to provide reasonable assurance that the licensing bases of all currently operating plants are sufficient to assure that operation is not inimical to the public health and safety.

6.3 Containment Systems

6.3.1 Safety Issues and Regulatory Requirements

The Commission regulations require that nuclear power plants be provided with an essentially leaktight containment as a barrier against uncontrolled release of radioactivity to the environment following accidents. More specifically, the regulations require that containment heat removal systems be designed, inspected, and tested in a manner intended to ensure their safety function and that containment atmosphere cleanup systems be designed, inspected, and tested in a manner intended to ensure their safety function. In addition, the regulations require that the containment be designed to (1) withstand post-accident temperature and pressure conditions without exceeding the design leak rate, (2) prevent fracture, (3) permit periodic integrated leakage testing, (4) permit periodic inspection and pressure testing of resilient seals, and (5) provide appropriate isolation valves.

6.3.2 Evolution of Current Licensing Basis

In order to demonstrate that containment designs are capable of withstanding post-accident temperature and pressure conditions without releasing excessive radioactivity, licensees and the staff have used mathematical models to establish and confirm acceptable containment performance. These models and the input assumptions are conservative and have demonstrated that containments are designed with substantial margin. As new information and research on containment design and post-accident energy release are obtained, such information is applied to the analytical methods as appropriate to ensure that adequate margins against excessive leakage are maintained.

For example, in the early 1970s, General Electric identified concerns regarding post-accident pool dynamic loads on BWR pressure suppression containments. The staff and BWR licensees performed significant reanalyses of containment performance based on this newly identified load phenomenon. The result of this effort was the formation of programs for modifications to the Mark I, II, and III BWR containment designs in order to reestablish the original containment design margins.

In the early 1980s, Westinghouse informed the staff that steam line break analyses may not have properly considered superheated steam blowdown conditions into the containment which could occur as the steam generator drys out. This information led to revised steam line break analyses by licensees which incorporated the new blowdown input. The new analyses confirmed that containment performance remains acceptable and appropriate margins are maintained.

Following the accident at Three Mile Island Unit 2 in March 1979, the NRC staff noted several concerns with regard to containment performance during the event that warranted improvement. One area of major focus concerned the capability to control combustible gas following accidents. Initially, the NRC staff required licensees to provide dedicated hydrogen penetrations to ensure the ability to employ hydrogen recombiners to reduce post-accident hydrogen concentration in the containment. This improvement was implemented as Item II.E.4.1 of the TMI Action Plan Clarification, NUREG-0737. However, the NRC staff also recognized that further research into combustible gas concerns was necessary. This led to substantial modifications to 10 CFR 50.44 in 1981 and 1985 wherein more stringent combustible gas control measures were specified for pressure suppression containment plants. Implementation of these requirements has improved combustible gas control capability.

The TMI-2 accident also pointed out the need for improvements in containment isolation dependability. New criteria in this regard were implemented as part of TMI-2 Action Plan, Item II.E.4.2, which required all licensees to evaluate their post-accident containment isolation capability against current criteria and make the necessary changes to improve its dependability.

Over the past few years, the NRC staff has undertaken research into severe accident effects on containments for all types of operating plants. To date, this program has pointed out weaknesses in the capability of BWR plants with Mark I pressure suppression containment designs to ensure adequate containment integrity under severe accident conditions. This has resulted in issuance of Generic Letter 89-16, which indicated the staff's intention to pursue plant-specific backfit procedures for a wetwell vent on all Mark I plants if the licensee does not voluntarily install the vent. Implementation of this improvement is currently proceeding.

Based on the continuous review of Appendix J leak rate test results, the NRC staff has periodically updated this rule to incorporate improved containment leak rate testing guidelines. One recent change was to permit use of the mass point method when conducting a Type A integrated leak rate test. Other revisions to Appendix J are currently pending and will provide further improvement in leak rate testing.

In addition, inservice inspection requirements for the containment structures and components are identified in Appendix J, 10 CFR Part 50. Section V.A in Appendix J requires a general inservice inspection of the accessible interior and exterior surfaces of the containment structures and components prior to any Type A test to uncover any evidence of structural deterioration that may affect either the containment structural integrity or leaktightness.

As part of the SEP effort the staff reviewed the isolations capability of lines that penetrate containment. Isolation provisions for lines that penetrate the primary containment maintain an essential leaklight barrier against the uncontrolled release of primary system coolant as result of postulated pipe breaks outside containment. The isolation function must be accomplished without endangering the performance of post-accident safety systems. Since the effort, the staff has implemented a number of regulatory initiatives that address the topic. TMI Action Plant item II.E.4.2 addressed actions relating to containment isolation system dependability, such as isolation of non-essential systems, diverse signals, and the reset of the containment isolation system (CIS) actuation. The TMI Action Plan also addressed procedural improvements. Type C testing required by Appendix J and by plant-specific TS periodically verify acceptable leak rates from containment. The NRC inspection program examines the valve arrangements and administrative controls on manual valves. Although not required to address this SEP issue, the IPE program will also include a review to identify specific vulnerabilities.

6.3.3 Conclusions

The staff review at the time of the initial licensing of a facility determined that the information provided by the applicant was sufficient for the staff to conclude that the containment design met the staff acceptance criteria and the intent of all applicable regulations. The staff has and will continue to obtain new information related to this subject area through a variety of sources such as updates to the FSAR, research reports, operating plant events and routine plant inspections. The staff reviews this information and, in the past, has required licensees to take actions to upgrade the plant to provide continuing assurance of adequate protection of the public health and safety. In addition, the staff will continue to review new information in this subject area and if the staff determines that new or different requirements are needed, the staff has the capability within the existing regulatory process to require additional analyses or plant modifications, as necessary, to ensure the continued health and safety of the public. In conclusion, the Commission concludes that the current regulatory process has and will continue to provide reasonable assurance that the licensing bases of all currently operating plants are sufficient to assure that operation is not inimical to the public health and safety.

6.4 Emergency Core Cooling Systems (ECCS)

6.4.1 Safety Issues and Regulatory Requirements

The Commission regulations require that nuclear power plants contain abundant emergency core cooling capability and specifies the specific safety functions for these systems. 10 CFR Part 50.46 and Appendix K to 10 CFR Part 50 establish the criteria and evaluation methods to be used by licensees and vendors to evaluate ECCS designs. The ECCS cooling performance must be evaluated using an acceptable model and must be evaluated for a number of postulated loss-of-coolant accidents of different sizes, locations, and other properties to ensure that the range of postulated loss-of-coolant accidents are considered.

6.4.2 Evolution of Current Licensing Basis

In June 1971, prior to establishing Part 50.46 or Appendix K, the Commission published interim acceptance criteria for ECCS designs by Westinghouse and General Electric reactor plants, and concluded that these criteria provide a reasonable assurance that ECCS will be effective in the unlikely event of a loss-of-coolant accident. However, research was under way at the time, and increased knowledge of heat transfer, fluid flow, and engineering disciplines important to ECCS analysis was anticipated.

Based on this research, modifications were made to the ECCS analysis guidelines. In December 1971, the NRC amended the interim criteria to add evaluation models for reactor designs by Babcock and Wilcox and Combustion Engineering. In January 1972, the AEC undertook an extensive rulemaking hearing. As a result of this proceeding, the Commission established a new Part 50.46 and Appendix K in January 1974, setting forth the acceptance criteria and the ECCS evaluation models in a final rulemaking. These regulations, which were enacted only after extensive rulemaking hearings, established the general approach that remains in use today. Between 1974 and 1976, extensive efforts were made to apply the requirements and criteria of Part 50.46 and Appendix K to all light water reactors then in operation. All plants subsequently licensed have been found to meet Part 50.46 and Appendix K.

In 1987, the Commission proposed modifications to the regulations because research, performed since the current rule was written, has shown that calculations performed using current methods and in accordance with the current requirements result in estimates of cooling system performance that are significantly more conservative than estimates based on the improved knowledge gained from this research.

The final rule incorporating new evaluation models was published in September 1988, but did not force facilities that had used previous models to perform new analyses. The Commission concluded at that time that existing Appendix K evaluation models should be permitted indefinitely. The Commission also believes that the decision to permit continued use of such models can and should be made at this time because it believes that both methods provide adequate protection of the public health and safety.

During the process of licensing, each applicant must submit in the FSAR sufficient information to describe the design bases for each ECCS subsystem, including its functional requirements, reliability requirements, protection from physical damage, and environmental conditions. Significant design parameters such as design flow rates, system temperatures, etc., along with piping and instrumentation diagrams, are routinely included.

Prior to granting an operating license, the staff reviews the described ECCS design against established acceptance criteria and concludes, in general, that the plant-specific design of the ECCS meets all necessary requirements and is acceptable.

The review process does not stop. The performance requirements of the ECCS are routinely evaluated during each plant refueling to ensure that operation during the subsequent cycle will be within the safety envelope of the plant design. In many instances, technical specification changes or license conditions are implemented to govern operation during the period of operation. In the extreme case, plant modification may be required to provide continued assurance of public health and safety.

Plant operating events also generate new information that may require operating plants to reanalyze the performance of the ECCS and, as necessary, make plant modifications. One such example was the lessons learned from the accident at TMI-2. Following this event, all operating reactors were required to reanalyze their plant-specific response to a range of small-break LOCAs. In some cases, these reanalyses resulted in plant modifications or changes in operating procedures being made. Another result of the TMI event was the requirement to install reactor head vents and to have operating procedures that describe how to use these vents in the event of certain postulated accidents. The net result was an overall improvement in the level of safety provided by the ECCS at operating nuclear power plants.

6.4.3 Conclusions

The staff review at the time of the initial licensing of a facility determined that the information provided by the applicant was sufficient for the staff to conclude that the ECCS designs met the staff acceptance criteria and the intent of all applicable regulations. The staff has and will continue to obtain new information related to this subject area through a variety of sources such as updates to the FSAR, research reports, operating plant events and routine plant inspections. The staff reviews this information and, in the past, has required licensees to take actions to upgrade the plant to provide continuing assurance of adequate protection of the public health and safety. In addition, the staff will continue to review new information in this subject area and if the staff determines that new or different requirements are needed, the staff has the capability within the existing regulatory process to require additional analyses or plant modifications, as necessary, to ensure the continued health and safety

of the public. In conclusion, the Commission concludes that the current regulatory process has and will continue to provide reasonable assurance that the
licensing bases of all currently operating plants are sufficient to assure that
operation is not inimical to the public health and safety.

6.5 Habitability Systems

6.5.1 Safety Issues and Regulatory Requirements

The Commission regulations require that control rooms at nuclear power plants
be provided with adequate radiation protection to permit access and occupancy
under accident conditions such that personnel do not receive radiation exposures in excess of 5-rem whole body or its equivalent to any part of the body
for the duration of the accident. Additional guidelines are contained in regulatory guides for assuring operator protection against both radioactivity and
toxic gas (e.g., chlorine) releases following postulated accidents.

6.5.2 Evolution of Current Licensing Basis

To satisfy the Commission requirements, all licensees have performed dose analyses using mathematical models to ensure that post-accident radiation levels
within the control room are within the required limits. Guidelines for conducting these analyses have remained essentially unchanged since the mid-1970s.
The assumptions used are considered to be conservative in order to account for
uncertainties in the actual radioactivity release mechanism following an
accident.

The accident at Three Mile Island Unit 2 in March 1979 pointed out potential
vulnerabilities in the capability of control room habitability systems, i.e.,
the control room ventilation system to ensure adequate radiation protection for
the operators. Therefore, Item III.D.3.4 of the TMI Action Plan Clarification,
NUREG-0737, was implemented at all operating plants. This item required
licensees to evaluate their control room habitability systems against the criteria of Standard Review Plan Section 6.4 and perform the necessary analyses of
toxic gas and radiation exposure to the operators in order to demonstrate compliance with these criteria. All plants provided responses to this issue and
made improvements in the control room ventilation systems, as appropriate.

The staff also recognized the need to conduct a longer term review of criteria
for ensuring control room operator protection and began a study in this regard
under Generic Issue 83 in the mid-1980s. This effort began with a survey of 12
nuclear power plants to determine what improvements had been made as part of
the NUREG-0737, Item III.D.3.4, implementation. Based on the results of the
survey, the staff determined that further guidance to improve control room habitability systems was necessary. This guidance is currently under development
and is intended to be issued in a generic letter to all licensees soon.

6.5.3 Conclusions

The staff review at the time of the initial licensing of a facility determined
that the information provided by the applicant was sufficient for the staff to
conclude that the control room habitability system met the staff acceptance
criteria and the intent of all applicable regulations. The staff has and will
continue to obtain new information related to this subject area through a variety

of sources such as updates to the FSAR, research reports, operating plant events and routine plant inspections. The staff reviews this information and, in the past, has required licensees to take actions to upgrade the plant to provide continuing assurance of adequate protection of the public health and safety. In addition, the staff will continue to review new information in this subject area and if the staff determines that new or different requirements are needed, the staff has the capability within the existing regulatory process to require additional analyses or plant modifications, as necessary, to ensure the continued health and safety of the public. In conclusion, the Commission concludes that the current regulatory process has and will continue to provide reasonable assurance that the licensing bases of all currently operating plants are sufficient to assure that operation is not inimical to the public health and safety.

6.6 Fission Product Removal and Control Systems

6.6.1 Safety Issues and Regulatory Requirements

The Commission regulations require that containment atmosphere cleanup systems be designed, inspected, and tested in a manner to ensure their safety function following postulated accidents.

6.6.2 Evolution of Current Licensing Basis

Staff guidance in the fission product removal and control area has changed little over the years. Most nuclear power plants are equipped with ventilation systems containing charcoal and high efficiency particulate air filters for fission product removal and prevention of unacceptable radiological releases during normal operation and post-accident conditions. Plants with filters have technical specifications that require surveillance and testing of those filters to ensure their continued satisfactory performance. PWR plants are also equipped with containment spray systems that provide both a post-accident heat removal and fission product control safety function in containment. As a means of controlling pH in the spray water, these plants have utilized a sodium hydroxide solution as a spray additive. Over the years, however, the staff recognized, through research of the post-accident source term, that a lower spray water pH (no lower than 7) was acceptable to ensure iodine retention and long-term corrosion control in ECCS systems. As a result, some PWR licensees have removed the sodium hydroxide addition system and replaced it with much simpler trisodium phosphate baskets placed directly in the containment sump in order to achieve necessary pH control.

In BWRs, blowdown of the reactor through the suppression pool results in some fission product removal following an accident. However, the staff had not previously credited this pathway in dose analyses. Staff review of recent analyses by General Electric resulted in a recognition of the suppression pool as a means of fission product control and led to a revision of the Standard Review Plan to credit an appropriate decontamination factor. Future BWRs will utilize this additional credit in post-accident dose analyses as may currently operating plants when proposing changes.

6.6.3 Conclusions

The staff review at the time of the initial licensing of a facility determined that the information provided by the applicant was sufficient for the staff to conclude that the fission product removal and control system met the staff acceptance criteria and the intent of all applicable regulations. The staff has and will continue to obtain new information related to this subject area through a variety of sources such as updates to the FSAR, research reports, operating plant events and routine plant inspections. The staff reviews this information and, in the past, has required licensees to take actions to upgrade the plant to provide continuing assurance of adequate protection of the public health and safety. In addition, the staff will continue to review new information in this subject area and if the staff determines that new or different requirements are needed, the staff has the capability within the existing regulatory process to require additional analyses or plant modifications, as necessary, to ensure the continued health and safety of the public. In conclusion, the Commission concludes that the current regulatory process has and will continue to provide reasonable assurance that the licensing bases of all currently operating plants are sufficient to assure that operation is not inimical to the public health and safety.

7 INSTRUMENTATION AND CONTROL SYSTEMS

7.1 Scope

The licensing bases and regulatory requirements for instrumentation and control (I&C) systems are discussed in the following sections.

The systems to be discussed in this section include the reactor trip system, engineered safety features actuation system, safe shutdown systems, and safety-related display systems. Remote shutdown systems are included in safe shutdown systems and post-accident monitoring and safety parameter display systems are included in safety-related display systems.

7.2 Development of Regulatory Requirements

Initially the regulatory requirements came from the need to develop highly reliable instrumentation and control systems to monitor and control the operation of nuclear reactors and other critical systems. In response to this need, concepts and methods such as the single failure criterion, failure mode and effects analysis, reliability, failure rates, sneak circuit analysis, redundancy, and diversity were developed and applied. In August 1968, these concepts and methods were originally collected into proposed IEEE Standard 279, "Criteria for Protection Systems for Nuclear Power Generating Stations," which was incorporated into 10 CFR 50.55a(h) in 1970. In addition, these concepts and methods were made part of the Commission regulations governing the design, fabrication, construction, installation, testing, and operation of these highly reliable instrumentation and control systems for nuclear reactors.

In 1974 and 1975, the staff went further in providing guidance by drafting and issuing criteria by which they would review the safety analysis reports (SARs) and other information submitted by licensees and applicants. The totality of these requirements has become the regulatory requirements that the licensees must address for their plant. This body of requirements is frequently revised and upgraded to take into account technological advances and lessons learned from operating experience.

The licensee, however, is authorized through 10 CFR 50.59 to make changes to the plant and its procedures and to conduct tests or experiments not described in the SAR without prior NRC approval unless the proposed change, test, or experiment involves changes to the technical specifications or introduces an unreviewed safety question. This body of requirements as it exists at the time application is made for an operating license and as reviewed and approved by the staff becomes the specific regulatory requirements for that plant.

7.3 Reactor Trip System

7.3.1 Safety Issues and Regulatory Requirements

The primary safety function of the reactor trip system (RTS) instrumentation is to monitor selected reactor and plant parameters related to nuclear power generation and transfer of the heat from that generation to the power conversion devices. When these parameters approach and exceed values deemed unsafe by analysis, the system shall initiate reactor shutdowns that shall promptly make the reactor core subcritical, i.e., stop the generation of nuclear power, by rapidly inserting control rods into the core or by other means of rapidly inserting enough negative reactivity into the core to make it subcritical and to keep the core subcritical.

The RTS instrumentation must be highly reliable, minimize false shutdowns, possess high availability, be automatically initiated, provide for manual initiation, and be designed so that the operators can easily and quickly determine the state of the plant. Applicable design requirements ensure that trip parameter monitoring channels and trip logic and actuation trains are redundant and independent; that all channels and trains meet the single failure criterion; that monitored parameters are sufficiently diverse; and that measuring instrumentation possesses adequate range, sensitivity, and accuracy and has adequate capability for test and calibration. These design require- ments ensure that parts and components are specified that meet plant-specific seismic and environmental requirements in accordance with IEEE Standard 344 and 10 CFR 50.49. Fabrication and installation requirements ensure that the system or subsystem is built of Class 1E parts and components, that it is fabricated and installed to meet plant-specific seismic and environmental requirements, and that quality control and quality assurance programs and procedures are used that meet the requirements of 10 CFR Part 50, Appendix B, and applicable IEEE and ANSI standards. Testing and operational requirements set forth in 10 CFR 50.36, 10 CFR Part 50 Appendices A and B, IEEE standards, and various regulatory guidance ensure that the system is adequately tested prior to and during operation and that the system is operated within the limits specified in the plant technical specifications.

7.3.2 Evolution of Current Licensing Basis

As plants became operational during the 1970s, operating experience indicated the need for improvements and changes in branch technical positions, technical specifications, regulatory guides, and IEEE standards. Changes at the nuclear power plants were also recommended in generic letters and bulletins. However, several major events occurred that caused major changes to be made to the licensing basis.

The Brown's Ferry fire in 1975 taught lessons about separating and protecting safety-related instrumentation, control, and power cabling. It also emphasized the importance of providing remote initiation capabilities for safety-related equipment that could be made independent of cabling and equipment in the cable spreading and main control rooms. Revision of the IEEE standard and the regula- tory guide on separation and independence as well as revisions to other IEEE standards relating to testing, qualification, and installation of safety-related equipment resulted from staff experience with this event. All plants licensed subsequently were reviewed by the staff to confirm that the protection system design precludes the use of components that are common to redundant channels, such as: actuation, reset, mode and test switches, common power supplies, or any other features that could compromise the independence of redundant channels. IEEE Std. 279 Sec. 4.6; IEEE Std. 384; Regulatory Guide 1.75; GDC-22; and SRPs 7.2 and 7.3 were used as acceptance criteria for these reviews.

The TMI-2 event in 1979 mandated many changes, which included significant revisions to operating procedures, incorporation of human factors concepts into the design and arrangement of instrumentation and controls on main control boards, monitoring of reactor vessel water level for BWRs and for PWRs, and new instrumentation to indicate reactor coolant sub-cooling margin for PWRs. The changes were implemented through generic letters and confirmatory orders.

The Salem ATWS (Anticipated Transients Without Scram) events in 1983 involved the only failure of a U. S. reactor to shut down on demand. The rapid intervention of the operators limited the consequences but the implications regarding shutdown reliability were significant and brought changes in operating procedures, reevaluation of on-line testing capability of the RTS, modifications to RTS breakers for B&W and Westinghouse plants, and changes to associated maintenance procedures. These improvements were requested by Generic Letter 83-28. Each licensee response was reviewed and approved by the staff. In 1984, the Commission issued 10 CFR 50.62, which added diverse and independent reactor trip systems to further improve reactor shutdown reliability and reduce the risk from potential occurrences of ATWS events. The NRC is presently reviewing and inspecting each plant to ensure that the systems have been installed properly.

7.4 Engineered Safety Features Actuation Systems

7.4.1 Scope

In this section are the actuating systems for typical ESF systems such as containment and reactor vessel isolation, emergency core cooling, containment heat removal, auxiliary feedwater, diesel generators, and standby gas treatment.

7.4.2 Safety Issues and Regulatory Requirements

The primary safety function of the engineered safety features actuation system (ESFAS) is to sense the need for, select, and initiate systems that take action to terminate or control and contain the effects and consequences of design basis accidents and operational occurrences.

As for the RTS instrumentation and logic, the ESFAS instrumentation, logic, and actuation equipment should also be highly reliable, minimize spurious actuations, possess high availability, be automatically initiated, provide for manual initiation of protective action from the control room, and be so designed that the operators can readily determine the status of the ESF systems and their actuating systems.

7.4.3 Evolution of Current Licensing Basis

The ESFAS and the RTS are very similar systems, the difference being principally in the systems controlled or actuated and the mission of those systems. The licensing basis for the two systems evolved in very much the same way. The same IEEE standards and the same regulatory guides apply to both systems. Some staff requirements apply to the ESFAS that do not apply to the RTS and vice versa; however, the basis for applying the requirements to the ESFAS and RTS subsystems is the same. Similarly, the modifications to the licensing bases for the ESFAS caused by the Brown's Ferry fire, TMI, Salem ATWS, the ATWS rule, and the feedback of operating experience are much the same for the ESFAS as they were for the RTS and are therefore not presented again.

7.5 Safety-Related Display Instrumentation

7.5.1 Scope

This section includes the post-accident monitoring instrumentation (PAM) and the safety parameter display system instrumentation with the normal safety-related display instrumentation.

7.5.2 Safety Issues and Regulatory Requirements

The primary safety function of the safety-related display instrumentation (SRDI) is to assist in meeting the Commission regulations by providing the capability to display the instantaneous values of the monitored plant operating parameters that provide the operators the information they need to form and update their assessment of the plant's operating status.

The primary safety function of the PAM is to provide the capability to monitor appropriate plant parameters during and after plant accidents and transients to assist the control room operators in preventing and mitigating the consequences of those events. The primary safety function of the safety parameter display system (SPDS) is to provide a concise display of critical plant variables to the control room operators to aid in rapidly and reliably determining the safety status of the plant.

As with the RTS, PAM instrumentation must be highly reliable, possess high availability, and be so designed that the operators can readily determine the status of the key variables. Applicable design requirements ensure that instrumentation channels are redundant and independent; that all channels meet the single failure criterion; that monitored parameters are sufficiently diverse; and that the measuring and indicating instrumentation possesses adequate range and sensitivity and has the capability for test and calibration. These design requirements also ensure that parts and components are specified that meet plant-specific seismic and environmental requirements in accordance with IEEE Standard 344 and 10 CFR 50.49. Fabrication and installation requirements ensure that the instrumentation is built of Class 1E parts and components, that it is fabricated and installed to meet plant-specific seismic and environmental requirements, and that quality control and quality assurance programs and procedures are used that meet the requirements of 10 CFR Part 50 Appendix B and applicable IEEE and ANSI standards. Testing and operational requirements set forth in 10 CFR 50.36, 10 CFR Part 50 Appendices A and B, IEEE standards, and various regulatory guides ensure that the instrumentation is adequately tested and that the instrumentation is operated within the limits specified in the plant technical specifications.

7.5.3 Evolution of Current Licensing Basis

The TMI-2 event mandated many changes in safety analysis philosophy, operating procedures, incorporation of human factors concepts into the design and arrangement of instrumentation and controls on the main control boards, monitoring of reactor vessel water level for BWRs and for PWRs, and reactor coolant sub-cooling margin for PWRs. Following the TMI-2 event, the NRC staff developed a comprehensive and integrated plan to improve safety at power reactors. As part of this plan, the Commission required the installation of improved post-accident monitoring instrumentation and SPDS. These improvements were intended to provide the operator with a broader range of information for accidents, including those beyond the design basis.

The SPDS and Regulatory Guide 1.97, Revision 2, were items identified in the TMI Action Plan (NUREG-0737). Additional clarification for implementation of these items was addressed in NUREG-0737, Supplement No. 1 via Generic Letter 82-33. The SPDS and the instrumentation in Regulatory Guide 1.97 were required for all operating plants, applicants for operating licenses, and holders of construction permits. The staff has reviewed almost all the submittals on conformance to Regulatory Guide 1.97 and SPDS. Generic Letter 89-06 was issued to all licensees for the purpose of certifying that the SPDS fully meets or will be modified to meet the requirements of NUREG-0737, Supplement 1.

7.6 Safe Shutdown and All Other Systems Required for Safety

7.6.1 Scope

This section includes systems and interlocks required for safe shutdown and safe operation of the reactor which were not included as part of either the reactor trip system, engineered safety features actuation system, or the safety-related display instrumentation. Examples include, for PWRs: residual heat removal, auxiliary feedwater, boration, interlocks, and radiation monitoring systems and remote shutdown facilities; for BWRS: reactor core isolation cooling, residual heat removal (shutdown cooling mode), standby liquid control, neutron monitoring (including rod block monitor), recirculation pump trip, interlocks, and radiation monitoring systems, low level set instrumentation, and remote shutdown facilities.

7.6.2 Safety Issues and Regulatory Requirements

The safety functions of the systems in this section vary with the system or equipment but, for the majority of these systems, it is preventive for the interlocks, boration, and SLCS systems and protective for the shutdown cooling systems and radiation monitoring systems.

In general, the instrumentation and logic systems in this section should meet the same safety criteria and regulatory requirements discussed in Section 7.4.1 for the ESFAS instrumentation and logic systems; however, some systems and portions of systems, particularly radiation monitoring systems and portions of interlock systems, may not be required to meet all the requirements for Class 1E systems. In addition, the Commission regulations require the provision for remote shutdown facilities that are located outside of the main control room and that meet the regulatory requirements.

7.6.3 Evolution of Current Licensing Basis

The licensing basis for these systems and equipment contains the same basic requirements as the RTS and ESFAS relating to system and equipment reliability; availability; redundancy; independence; ability to meet the single failure criterion; provision of adequate range, sensitivity, and accuracy in sensing and monitoring equipment; and provision of capability for test and calibration of the systems and equipment to which these requirements apply.

Modifications to the licensing basis for the safe shutdown and all other systems required for safety that were found necessary by experience gained from the Brown's Ferry fire, TMI, Salem ATWS, the ATWS rule, and the feedback of operating experience that updates it are the same for the requirements applicable to these systems and equipment as they are for the RTS and ESFAS systems. These were previously discussed in Sections 7.2.2 and 7.3.2 and will not be discussed further.

7.7 Control Systems

7.7.1 Scope

This section includes those control systems used for normal operation that are not relied upon to perform safety functions following anticipated operational occurrences or accidents but that control plant processes having a significant impact on plant safety. Examples include the reactivity control systems; the reactor coolant pressure, temperature, flow, and inventory controls; the secondary system pressure and flow controls; and the environmental control systems for safety-related instruments and instrument sensing lines.

7.7.2 Safety Issues and Regulatory Requirements

The licensing of earlier plants in the control system area usually encompassed the review of the interaction of the control systems with the safety systems that have been discussed in the previous sections. This review was performed to ensure that no interactions existed that would prevent or inhibit the safety system from performing its intended safety function.

7.7.3 Evolution of Current Licensing Basis

As the licensing process evolved, the review of the control system mentioned above became somewhat more detailed. In addition, the environmental control systems were added to the list of significant control systems and a regulatory guide that detailed the review bases for this system was published. During the later licensing years, plant-specific studies were performed to determine the effects of high energy line breaks on control systems, the effects of the loss of power to control systems used to shut the plant down in a normal manner, and the results of multiple control system failures on the existing safety analysis. Accordingly, requirements and criteria for the review of these control systems have been included in the SRP, and they have been made a part of the licensing bases.

A generic study was undertaken (USI A-47) that led to the conclusion that some modifications to plants should be made and that the failure of some control systems would have an impact on the safety analysis and, therefore, surveillance of these systems should be included in the technical specifications along with the safety systems mentioned in the previous sections. For example, Generic Letter 89-19 requested all reactor licensees to install, if not already present, overfill protection instrumentation. All responses to the generic letters as well as all changes will be reviewed by the staff.

7.8 General Conclusions

The staff review at the time of the initial licensing of a facility determined that the information provided by the applicant was sufficient for the staff to conclude that instrumentation and control systems, including the reactor trip, engineered safety features actuation, safe shutdown and safety-related display systems, met the staff acceptance criteria and the intent of all applicable regulations. The staff has and will continue to obtain new information related to this subject area through a variety of sources such as updates to the FSAR, research reports, operating plant events and routine plant inspections. The staff reviews this information and, in the past, has required licensees to take actions to upgrade the plant to provide continuing assurance of adequate protection of the public health and safety. In addition, the staff will continue to review new information in this subject area and if the staff determines that new or different requirements are needed, the staff has the capability within the existing regulatory process to require additional analyses or plant modifications, as necessary, to ensure the continued health and safety of the public. In conclusion, the Commission concludes that the current regulatory process has and will continue to provide reasonable assurance that the licensing bases of all currently operating plants are sufficient to assure that operation is not inimical to the public health and safety.

8 ELECTRIC POWER

8.1 Scope

Electric power systems power safety-related equipment that is necessary for mitigating the consequences of design-basis accidents and for bringing the plant to a safe condition and maintaining it in that condition. Electric power systems comprise an offsite power system and an onsite power system. These two systems will be discussed jointly in this chapter since their licensing basis is often contained in common regulatory requirements.

8.2 Safety Issues and Regulatory Requirements

Commission regulations establish the basic criteria to which the offsite power system and the onsite power system must be designed. These regulations require that each system (offsite and onsite) have sufficient capacity and capability by itself to support vital functions necessary to respond to operational occurrences and mitigate the consequences of design-basis accidents. In addition, the onsite power system must be able to withstand a single failure and the offsite power system must have two power circuits designed and located so as to minimize to the extent practical their simultaneous failure.

The regulations also require that the electric power systems be designed to permit appropriate periodic testing and inspection and that all operating plants have the capability to withstand and recover from a station blackout (loss of all ac power). These regulations also apply to portions of the electric power systems insofar as they provide general requirements for safety systems or provide requirements for systems that interface with the electric power systems.

8.3 Evolution of Current Licensing Basis

Commission regulations published in February 1971 constitute the primary licensing basis for the electric power systems. More recently, these regulations have been supplemented with the requirements in 10 CFR 50.63, published in June 1988, that require all plants be able to withstand and recover from a station blackout (loss of all ac power). The station blackout rule illustrates how the regulatory process functions to modify the licensing basis in the electric power systems area when a need is identified.

As operating experience was accumulated from license event reports (LERs), diesel generator failure reports, and feedback from the regions, a concern arose that the offsite and onsite emergency ac power systems might be less reliable than originally anticipated, even designs that met the requirements of the Commission regulations. Some operating plants had experienced a total loss of offsite power, and operating experience with onsite emergency power systems included many instances of diesel generators failing to start. In a few cases, there was even a complete loss of both the offsite and the onsite ac power systems, although ac power was restored in a short time without any serious consequences. In 1975, the results of the Reactor Safety Study (WASH-1400) showed that station blackout could be an important contributor to the total risk from nuclear power plant accidents. Although this total risk was found to be small, the relative importance of the station blackout accident was established. Subsequently, the Commission designated the issue of station blackout as an unresolved safety issue (USI), and initiated studies to determine if additional safety requirements were needed.

As a result of the station blackout studies, a proposed rule was published for comment in the Federal Register in March 1986. The final rule was published in June 1988. Concurrent with the development of this regulatory guidance, the Nuclear Management and Resources Council (NUMARC) also developed detailed guidelines and procedures for assessing station blackout capabilities at light-water reactors (NUMARC 87-00). The staff reviewed and approved NUMARC 87-00 and found the guidance therein acceptable for implementing the station blackout rule. The purpose of the effort in developing the NUMARC 87-00 guidelines and the staff's cooperation with this effort was to iron out differences and misunderstandings in advance and to establish acceptable approaches to various station blackout issues for all utilities in responding to the rule.

In April 1988, the staff received all licensee responses to the station blackout rule. The staff is presently reviewing these submittals and will issue safety evaluation reports (SERs) for each plant when the review is completed. It is expected that all licensees will have implemented all required modifications and procedure changes within 3 years. Although final plans for inspection have not been completed, it is likely that an audit inspection will be performed at some plants to monitor licensee implementation efforts.

In addition to rule changes, the staff has employed other less rigorous methods to make improvements in the electric power systems area when a need is identified. These include the use of generic letters, bulletins, revision to regulatory guides, creation of new regulatory guides, modifications to the standard review plan, and, more recently, cooperation with the nuclear power industry in the development of industry-sponsored guidance documents.

Generic Safety Issues B-23 and B-48 on degraded grid voltages and station electric distribution system voltages are examples of the use of generic letters to implement improvements to electric power systems. Events at Millstone and Arkansas Nuclear One power plants raised a concern that the offsite power systems required by Commission regulations may not satisfy the capability requirements of the criteria because they may not always provide adequate voltages to operate safety-related loads. This could cause loss of or damage to redundant safety systems during an event. As a result, the staff issued generic letters to all power reactor licensees in June 1977 and August 1979 requesting that they analyze their electric distribution systems for adequate voltages and describe to the staff the modifications to upgrade the protection of electrical relaying that separates the offsite power system from the safety loads when voltage levels are insufficient to operate these loads. These guidelines were later incorporated into a new branch technical position (PSB-1) in the standard review plan in order to ensure they are consistently applied to new plant license applications. As of today, all operating plants have submitted and received approval for plant modifications that implemented a second level of voltage protection for their safety-related electrical buses.

NRC Bulletin 88-10 on nonconforming molded-case circuit breakers is an example of how the staff has used a bulletin to require that licensees take some action to verify that their electrical system is in conformance with Commission requirements. Here again it was a question of whether the plant electric power systems were continuing to meet the capability requirements specified in the applicable Commission regulations. The staff found that at some plants circuit breakers supplied by a particular vendor were refurbished, not new as indicated by the supplier, and that several breakers did not meet required performance specifications. If these breakers were used in safety-related circuits, the reliable functioning of the circuit could not be assured. As a result, the staff issued Bulletin 88-10 in November 1988 requesting that all licensees verify traceability of certain circuit breakers used in safety systems and test those breakers where traceability to the original manufacturer could not be shown and report the results to the Commission. The staff is reviewing the licensee responses to determine if the licensee has implemented the actions contained in the bulletin. If the staff determines that the licensee has implemented the actions contained in the bulletin, no further staff action will be performed. If a licensee proposes alternative actions, the staff will handle these proposals on a case-by-case basis.

Generic Safety Issue (GSI) B-56 on diesel generator reliability improvement serves as an example of the use of revised regulatory guides and industry-sponsored guidance documents to implement improvements to electric power safety systems. GSI B-56 was initiated as a response to the lower than expected reliability of diesel generators as emergency power sources in the onsite power systems. It is related to the station blackout issue in that it is one of the primary sources of unreliability in the total loss of ac power event. As a result, the station blackout regulatory guidance called for a reliability program at nuclear power plants designed to monitor and maintain the reliability of the diesel generators and improve the reliability if an acceptable level is not achieved. Specific guidance to the utilities on how to implement such a program is being provided under the GSI B-56 resolution in the form of a revision to Regulatory Guide 1.9, which will reference a NUMARC document for the diesel generator reliability program recommendations. The NUMARC document (NUMARC 87-00, Appendix D) was produced by the nuclear power industry with input from the NRC staff (as described above) in the discussion on station blackout. The resolution of GSI B-56 will be complete when, consistent with the requirements of the station blackout rule, each licensee implements an emergency diesel generator reliability program to enhance the reliability of the onsite diesel generators.

Problems in the electric power systems, such as those discussed above, are identified by the staff on an ongoing basis through the review of LERs and other licensee notification requirements and through the staff's various license review and inspection activities. In addition to the vehicles identified above for making changes to the licensing basis for electric power systems, the staff often issues information notices to notify licensees of problems found in the electric systems at some plants. Although the notices do not require licensees to take any action, they serve to quickly advise licensees of problems that may exist in their plants, while the staff determines what, if any, additional action is warranted.

8.4 Conclusions

The staff review at the time of the initial licensing of a facility determined that the information provided by the applicant was sufficient for the staff to conclude that the design of the facility's electrical power systems met the staff acceptance criteria and the intent of all applicable regulations. The staff has and will continue to obtain new information related to this subject area through a variety of sources such as updates to the FSAR, research reports, operating plant events and routine plant inspections. The staff reviews this information and, in the past, has required licensees to take actions to upgrade the plant to provide continuing assurance of adequate protection of the public health and safety. For example, the Commission recently issued 10 CFR 50.63 to require electrical system upgrades. In addition, the staff will continue to review new information in this subject area and if the staff determines that new or different requirements are needed, the staff has the capability within the existing regulatory process to require additional analyses or plant modifications, as necessary, to ensure the continued health and safety of the public. In conclusion, the Commission concludes that the current regulatory process has and will continue to provide reasonable assurance that the licensing bases of all currently operating plants are sufficient to assure that operation is not inimical to the public health and safety.

9 AUXILIARY SYSTEMS

9.1 Scope

Auxiliary systems are those secondary systems provided to support operation and function of primary engineered safety features (ESFs); other systems not directly related to safe reactor operation and safe shutdown are also included. Their primary function is to remove heat from essential components (e.g., cooling water and ventilation systems) or provide motive power (e.g., compressed air) to equipment needed for safe reactor operation and postaccident shutdown. Support systems include cooling water systems (e.g., station service water, reactor auxiliaries cooling water, and the ultimate heat sink); compressed air systems; heating, ventilation, and air conditioning (HVAC) systems for various plant areas; and diesel generator auxiliaries (e.g., fuel oil, cooling water, lubrication, and combustion air systems). Other auxiliary systems not directly related to safe reactor operation and shutdown include new and spent fuel storage and handling systems, process sampling system, equipment and floor drainage system, fire protection system, and communication and lighting systems. In addition, the chemical and volume control system that provides normal reactor coolant system inventory in pressurized-water reactor (PWR) plants and the standby liquid control system that provides an emergency backup means of reactivity control in BWR plants are also within the scope of the auxiliary systems.

9.2 Safety Issues and Regulatory Requirements

Commission regulations require that cooling water systems supporting primary ESF systems be designed, tested, and inspected in a manner intended to ensure their safety function and that ESF systems be compatible with environmental conditions, which includes HVAC systems relied on to provide proper ESF equipment operating conditions. These regulations also require that spent fuel storage and handling systems be designed with features that ensure a safe spent fuel storage facility and that nuclear power plants be designed to minimize the probability of fires and have fire protection features to minimize the adverse effects of fires. Specific additional fire protection requirements are contained in 10 CFR 50.48 and Appendix R to 10 CFR Part 50.

9.3 Evolution of Current Licensing Basis

The licensing basis for auxiliary systems has evolved as reactor events and generic studies by the NRC staff provide new information that is determined to be helpful in improving the performance of auxiliary systems. The process of evaluating operating experience and assessing plant data to determine the need for additional actions is a continuous one.

One important source of operating experience information is the reports on events and equipment failures prepared by all licensees in accordance with the requirements of 10 CFR 50.72 and 10 CFR 50.73. The NRC Office for Analysis and Evaluation of Operational Data (AEOD) reviews this information and develops recommendations for action. Examples of this with regard to auxiliary systems are discussed below.

Another means of identifying the need for further actions is through the process of identifying, prioritizing, and evaluating generic issues when potential safety concerns arise that require longer term study. Examples of the generic issue process on auxiliary systems are also discussed below.

In 1980, the NRC staff became aware through reported events of fouling of service water systems and the resulting degradation in system performance caused by biological organisms. This resulted in the issuance of IE Bulletin 81-03, "Flow Blockage of Cooling Water to Safety-Related Components by Corbiocula Sp. (Asiatic Clam) and Mytilus Sp. (Mussel)," dated April 10, 1981. Licensees were requested to assess the potential for biofouling at their sites and implement appropriate monitoring or corrective actions. Subsequent service water system problems were also identified and generically communicated in IE Information Notice (IN) 85-30, "Microbiologically Induced Corrosion of Containment Service Water System" (April 19, 1985); IN 86-11, "Inadequate Service Water Protection Against Core Melt Frequency" (February 25, 1986); and IN 86-96, "Heat Exchanger Fouling Can Cause Inadequate Operability of Service Water Systems" (November 20, 1986).

In response to service water system degradation problems, several generic issues (GIs), but primarily GI-51, "Proposed Requirements for Improving Open Cycle Service Water Systems," were initiated to study the need for further recommendations for improving service water system performance. In addition, as part of its responsibility to evaluate operational data, AEOD undertook a study of service water system problems. The AEOD findings were eventually published in "Operational Experience Feedback Report - Service Water System Failures and Degradations," NUREG-1275, Volume 3, dated November 1988.

The AEOD report and GI-51 resolution led to development and issuance of Generic Letter (GL) 89-13, "Service Water System Problems Affecting Safety-Related Equipment," which recommended additional performance monitoring and design verifications in order to ensure the safety function of the service water system. All plants were required to respond to GL 89-13 by indicating their plans for accomplishing the NRC staff's intent to improve service water system performance. Through the inspection program, the NRC staff is performing audits of the implementation of the actions identified by the licensees in response to the generic letter and will assess the adequacy of the licensee's actions.

A similar process was followed during the early 1980s to correct reported failures and problems with degradation in the instrument air systems. Parallel NRC staff evaluations were conducted under GI-43, "Air System Reliability," and in AEOD, which resulted in publication of "Operational Experience Feedback Report - Air System Problems," NUREG-1275, Volume 2, dated December 1987. The GI-43 resolution and the AEOD report led to development and issuance of GL 88-14, "Instrument Air Supply System Problems Affecting Safety-Related Equipment," which requested that licensees perform a design-basis verification of their instrument air systems and make the necessary improvements to ensure its proper function. All licensees were required to respond to GL 88-14 indicating that they had accomplished the recommended actions. Within the framework of the inspection program, the NRC staff is performing audit inspections of instrument air systems to assess the adequacy of the licensees' actions for improving the systems' performance.

After the TMI-2 accident, it became apparent from analysis of the event that additional means were necessary to ensure prompt and accurate postaccident sampling of the containment environment and reactor coolant conditions in order to get the information needed to manage recovery. On the basis of experience gained from the TMI-2 accident, Item II.B.3, "Post-Accident Sampling Capability," was incorporated in NUREG-0737. All nuclear power plants were required to implement postaccident sampling capabilities in accordance with these guidelines.

Because of the safety significance of the fire at Browns Ferry Unit 1 in 1975, the staff undertook a comprehensive effort to develop more specific criteria to improve fire safety. This effort resulted in issuance of various staff positions and guidance in 1976, which included Branch Technical Position (BTP) Auxiliary and Power Conversion System Branch (APCSB) 9.5-1 and Appendix A to BTP APCSB 9.5-1. The Commission eventually codified fire protection requirements when it issued 10 CFR 50.48, "Fire Protection," and Appendix R to 10 CFR Part 50, dated November 19, 1980. All licensees whose plants had been licensed to operate before January 1, 1979, were required to compare their plant fire protection features to the new criteria and make the necessary modifications. Plants not licensed in 1979 or later were reviewed against similar criteria as part of the normal prelicensing review. Completion of these actions has resulted in sub-stantial improvement in fire protection and post-fire safe shutdown capability in all plants.

With the recognition in the mid-1970s that spent fuel from commercial nuclear power plants would not be reprocessed, it became apparent that much greater quantities of spent fuel would be stored in onsite spent fuel pools. This led to the development of additional guidance for ensuring safe spent fuel storage when license amendments were requested for expanding capacity in spent fuel pools. This guidance was issued to all nuclear power reactor licensees by a generic letter, dated April 14, 1978. This guidance continues to serve as a basis for ensuring safe onsite storage of spent fuel. Subsequent generic concerns regarding such spent fuel storage safety were evaluated by the NRC staff under Generic Issue 82, "Beyond Design Basis Accidents in Spent Fuel Pools." This effort led to the determination that additional criteria beyond these currently established for ensuring safe spent fuel storage were not neces-sary. In addition, the requirements of 10 CFR Part 72 must be satisfied if a licensee proposes to store spent fuel in an independent storage facility separate from the spent fuel pool itself.

Concerns with regard to the safe handling of heavy loads at nuclear power plants were the subject of a generic study under Generic Technical Activity A-36 during the late 1970s. This study resulted in publication of NUREG-0612, "Control of Heavy Loads at Nuclear Power Plants," dated July 1980, and issuance of a generic letter, dated December 22, 1980. The generic letter requested that, as stated in NUREG-0612, all licensees implement improvements to procedures, training, identification of safe load paths, and crane and lifting device maintenance and testing, in order to reduce the probability of a heavy load drop near spent fuel or safety-related equipment that could lead to an unacceptable release of radio-activity. The staff reviewed and provided a safety evaluation of each licensee's proposed actions to handle heavy loads more safely. Through the inspection pro-gram, the NRC staff audited licensee-identified actions to satisfy the concerns identified in NUREG-0612. Following these reviews, the staff undertook a pilot program to assess the need for implementation of additional NUREG-0612 guide-lines. On the basis of the pilot program, the staff determined that further actions recommended in NUREG-0612, which included the installation of single-failure-proof cranes and performance of load drop analyses, were not necessary. This conclusion was described in Generic Letter 85-11, dated June 28, 1985, wherein it was determined that actions already completed by licensees have satisfactorily reduced the probability of unacceptable heavy load drops.

Since the TMI-2 accident, the staff has begun in a systematic manner to review the capability of nuclear power plants to cope with beyond-design-basis (severe) accidents. This effort has relied largely on probabilistic risk assessment techniques. A major objective of these reviews was to identify potential plant vulnerabilities and take corrective actions accordingly. These reviews show that auxiliary and support systems can be dominant contributors to risk, and attention to their continued proper operation is important to plant safety. Future licensee activities requested by the staff as part of the Individual Plant Examination of internally initiated events (IPE) and the Individual Plant Examination of External Events (IPEEE) will include a focus on auxiliary systems and their contribution to plant safety.

9.4 Conclusions

The staff review at the time of the initial licensing of a facility determined that the information provided by the applicant was sufficient for the staff to conclude that the facility's auxiliary systems met the staff acceptance criteria and the intent of all applicable regulations. The staff has and will continue to obtain new information related to this subject area through a variety of sources such as updates to the FSAR, research reports, operating plant events and routine plant inspections. The staff reviews this information and, in the past, has required licensees to take actions to upgrade the plant to provide continuing assurance of adequate protection of the public health and safety. Examples of instances where staff review and examination has led to issuance of new criteria and implementation of improvements to auxiliary systems include service water system upgrades (GL 89-13), instrument air system upgrades (GL 88-14), post accident sampling in public upgrades (NUREG 0737, Item II.B.3), fire protection upgrades (10 CFR 50.48 and Appendix R to 10 CFR Part 50), spent fuel storage upgrades (GL of April 14, 1978), and heavy load capability upgrades (NUREG-0612). In addition, the staff will continue to review new information in this subject area and if the staff determines that new or different requirements are needed, the staff has the capability within the existing regulatory process to require additional analyses or plant modifications, as necessary, to ensure the continued health and safety of the public. In conclusion, the Commission concludes that the current regulatory process has and will continue to provide reasonable assurance that the licensing bases of all currently operating plants are sufficient to assure that operation is not inimical to the public health and safety.

10 STEAM AND POWER CONVERSION SYSTEM

10.1 Scope

The steam and power conversion system consists of those balance-of-plant systems necessary to provide feedwater to the reactor in boiling-water reactors (BWRs) and steam generators in pressurized-water reactors (PWRs) in order to produce the main steam supply to the turbine for generating power as part of the normal operating function of the nuclear power plant. With the exception of system piping interfaces to the primary (in BWRs) and secondary pressure boundary (in PWRs), these systems have no safety function and are not relied on to ensure a safe postaccident shutdown with one exception. The auxiliary feedwater system in PWR plants has an important postaccident and transient decay heat removal safety function, as is discussed below.

10.2 Safety Issues and Regulatory Requirements

Commission regulations require that the reactor coolant pressure boundary have an extremely low probability of abnormal leakage, of rapidly propagating failure, and of gross rupture under operating, maintenance, testing, and postulated accident conditions. These requirements pertain to the steam and power conversion system in PWRs because control of the secondary water chemistry and inservice inspection to technical specifications limits are essential to ensuring steam generator tube integrity and preventing unacceptable primary coolant leakage into the secondary (steam) system.

In addition, the regulations require that nuclear power plants have a system to remove residual heat following accidents and transients and specify design requirements for the system. In PWRs, the auxiliary feedwater system provides this function for most events, except for postulated large reactor coolant piping failures.

10.3 Evolution of Current Licensing Basis

The 1979 accident at Three Mile Island, Unit 2 (TMI-2) heightened the NRC staff's awareness of the importance of the postaccident decay heat removal safety function provided by the auxiliary feedwater system. Improper isolation of the auxiliary feedwater flow path at TMI-2 delayed the initiation of decay heat removal through the steam generators. This resulted in implementation of Items II.E.1.1 and II.E.1.2 of NUREG-0737, which required upgrades to the auxiliary feedwater system in all PWR plants to improve its reliability. The specific improvements were identified in NUREG-0611 and NUREG-0635 and included changes in system design, including initiation and flow indication, operating procedures, and technical specifications. The staff reviewed licensee responses to this item and wrote a safety evaluation for each plant.

Despite the improvements obtained by this effort, concern with auxiliary feedwater system reliability remained. This concern grew out of the review of auxiliary feedwater system reliability studies and continued failures noted from operating experience data reviews. The specific concern was that the availability of an auxiliary feedwater system with two pumps was not sufficient to ensure the secondary decay heat removal safety function when compared to that of a three-pump system. This concern was amplified by the loss of all feedwater event at Davis-Besse in June 1985. The staff pursued the issue under Generic Issue (GI) 124 where those few plants with just two auxiliary feedwater pumps were evaluated to determine the need to make further hardware changes to improve their auxiliary feedwater system reliability.

GI 124 was ultimately resolved with a requirement that two-pump auxiliary feedwater system plants backfit a third means of removing decay heat through the steam generators. Those few licensees affected by this decision have committed to implement this additional improvement for ensuring the auxiliary feedwater system safety function.

In addition, although the remaining portions of the power and conversion system do not perform a direct function in ensuring postaccident plant safety, events involving balance-of-plant systems have made the staff recognize that certain improvements were necessary in order to ensure the safety function of interfacing systems or reduce the likelihood of unanticipated plant trips. Two such areas of improvement include preventing erosion/corrosion and preventing waterhammer.

In the 1970s, waterhammer events in main feedwater systems at several PWR plants, including Indian Point Unit 2, Calvert Cliffs, and others, demonstrated the need for hardware improvements in order to reduce the chance of breaching the secondary side of the steam generator. The staff evaluated this issue under Unresolved Safety Issue (USI) A-1. As a result of this effort, PWRs have installed J-tubes on the feedwater ring header within the steam generator to reduce the likelihood of steam void formation in the feedwater line and potential waterhammer from collapse of the steam bubble on auxiliary feedwater system initiation. These actions have been effective at reducing the probability of damaging waterhammer.

The feedwater line break event at Surry Unit 2 in December 1986 pointed out the adverse consequences to plant safety from unplanned reactor trips caused by to balance-of-plant failures and to personnel from high energy system steam releases. As a result, the staff issued Bulletin 87-01, which requested that all licensees examine plant piping for wall thinning and take corrective action as necessary. A subsequent audit of licensee actions in response to Bulletin 87-01 indicated that continued programs to monitor for future erosion/corrosion were not in place at the plants. Therefore, the staff issued Generic Letter 89-08, which requested that licensees implement a continuous monitoring program to detect unacceptable pipe wall thinning and certify that the program is in place. These programs provide the necessary assurance against the type of severe wall thinning event that challenges plant safety systems. Through the inspection program, the staff audits licensee actions to ensure that adequate implementation has been undertaken.

As a result of steam generator tube degradation and leakage problems at many PWR plants in the 1970s, PWR licensees, nuclear steam supply system (NSSS) vendors, and the NRC staff initiated studies to improve steam generator tube integrity. One major early outcome of these studies was a recognition that typical secondary water chemistry programs that included sodium phosphate were potentially contributing to the tube degradation being experienced. As a result, NSSS vendors recommended a change to an all volatile treatment (AVT), secondary water chemistry program utilizing ammonia and hydrazine. Licensees have adopted this change, and subsequent operating experience indicates that it has been effective in improving steam generator tube integrity. Steam generator tube integrity was also designated an unresolved safety issue (USI) by the NRC staff in 1978 and Task Action Plans (TAPs) A-3, A-4, and A-5 were established to evaluate the safety significance of degradation in Westinghouse, Combustion Engineering, and Babcock and Wilcox steam generators, respectively. These studies were later combined into one effort because many of the problems being experienced at these plants were similar. The staff prepared a draft USI report regarding this issue, which primarily considered corrosion-related failure mechanisms, including the "denting" mechanism, since those failures were the main concern during the period when most of the technical studies were performed.

In May 1982, subsequent to the Ginna steam generator tube rupture (SGTR) event, the staff initiated an integrated program to consider the lessons learned from the Ginna SGTR event, and from the three previous domestic SGTR events. The staff also considered the recommendations in the draft USI report. The objective of the integrated program was to resolve USIs A-3, A-4, and A-5 and determine the need for further requirements to improve steam generator tube integrity.

Concurrent with the completion of the staff study under USIs A-3, A-4, and A-5, in 1985, the NRC staff issued Generic Letter 85-02, which requested that all PWR licensees describe their programs, including secondary water chemistry control, for ensuring steam generator tube integrity. The NRC staff reviewed these programs and accepted them with necessary changes made by licensees.

The results of the NRC staff integrated program for resolution of USIs A-3, A-4, and A-5 were ultimately documented in NUREG-0844, "NRC Integrated Program for the Resolution of Unresolved Safety Issues A-3, A-4, and A-5 Regarding Steam Generator Tube Integrity," September 1988. In NUREG-0844, the NRC staff concluded that sufficient regulatory requirements were in place, in conjunction with industry initiatives, to ensure that SGTRs do not contribute significantly to nuclear power plant risk, and thus, no further regulatory requirements were necessary.

10.4 Conclusions

The staff review at the time of the initial licensing of a facility determined that the information provided by the applicant was sufficient for the staff to conclude that those portions of the steam and power conversion system performing essential safety functions met the staff acceptance criteria and the intent of all applicable regulations. The staff has and will continue to obtain new information related to this subject area through a variety of sources such as updates to the FSAR, research reports, operating plant events and routine plant inspections. The staff reviews this information and, in the past, has required licensees to take actions to upgrade the plant to provide continuing assurance of adequate protection of the public health and safety. The TMI-2 accident led to significant improvements in the availability and reliability of the auxiliary feedwater system, which is the portion of the steam and power conversion system providing a decay heat removal safety function in PWRs. Events at other plants in balance-of-plant systems led to improvements to reduce the likelihood of damaging waterhammer and unanticipated plant trips due to errosion/corrosion of piping. In addition, the staff will continue to review new information in this subject area and if the staff determines that new or different requirements are needed, the staff has the capability within the existing regulatory process to require additional analyses or plant modifications, as necessary, to ensure the continued health and safety of the public. In conclusion, the Commission concludes that the current regulatory process has and will continue to provide reasonable assurance that the licensing bases of all currently operating plants are sufficient to assure that operation is not inimical to the public health and safety.

11 RADIOACTIVE WASTE MANAGEMENT SYSTEMS

11.1 Scope

Radioactive waste management systems are provided to control releases of radioactive materials to the environment in liquid and airborne effluents and to handle radioactive solid wastes produced during normal reactor operation. Process and effluent radiological monitoring and sampling systems are provided for monitoring effluent discharge paths for radioactivity that may be released from normal operations and from postulated accidents.

Radioactive liquid and solid waste management systems are relatively independent of the type of plant; however, radioactive gaseous waste management systems at boiling-water reactors (BWRs) differ significantly from those at pressurized water reactors (PWRs).

11.2 Safety Issues and Regulatory Requirements

Commission regulations limit and otherwise govern the radioactivity in effluents released to unrestricted areas for the purpose of providing protection against the hazards of radiation from normal plant operation. These regulations (1) provide requirements regarding the characteristics of radioactive waste prepared and packaged for transfer to offsite disposal sites, (2) provide design objectives for equipment to control releases of radioactive materials in effluents, and (3) require technical specifications to keep releases of radioactive materials to unrestricted areas during normal operations as low as is reasonably achievable.

Furthermore, Commission regulations require that (1) the plant design include means to suitably control the release of radioactive materials in gaseous and liquid effluents and to handle radioactive solid wastes produced during normal operations, including anticipated operational occurrences; (2) radioactive waste systems be designed to ensure adequate safety under normal and postulated accident conditions; and (3) means be provided for monitoring effluent discharge paths for radioactivity that may be released from normal operations, including anticipated operational occurrences, and from postulated accidents.

11.3 Evolution of Current Licensing Basis

Significant changes to the regulations governing radioactive waste management systems occurred with the establishment of Appendix I to 10 CFR Part 50 in 1975 and 10 CFR 20.311, and 10 CFR Part 61 in 1982. Appendix I was issued because the NRC recognized that specific numerical criteria were necessary to ensure that licensees were maintaining radioactivity levels within normal effluent releases to as low as reasonably achievable limits as required by 10 CFR Part 20.

The technical specifications established under 10 CFR 50.36a ("Appendix I technical specifications") for all plants are intended to ensure that radioactive waste processing operations are conducted within specific limits. These technical specifications provide limiting conditions for operation and surveillance requirements regarding (1) operation of the liquid and gaseous radwaste treatment systems, (2) radioactive materials in liquid and gaseous effluents, (3) offsite doses due to radioactive materials in liquid and gaseous effluents, (4) total offsite doses, (5) radiological environmental monitoring, (6) content of liquid and gaseous waste storage tanks, (7) explosive mixtures in gaseous radwastes management systems, and (8) processing of solid radioactive wastes.

Subsequent experience and problems with the acceptability of solid waste packages from commercial nuclear power plants intended for burial at licensed offsite facilities resulted in issuance of 10 CFR Part 20 requirements to more closely control transfers of radioactive waste intended for disposal at a licensed land disposal facility. These problems included excessive amounts of water in solid waste packages and overly rapid deterioration of the waste form itself. Under 10 CFR 20.311, licensees are required to (1) prepare all solid wastes so that they can be classified according to 10 CFR 61.55, (2) meet the waste characteristic requirements of 10 CFR 61.56, and (3) conduct a quality control program to ensure compliance with 10 CFR 61.55 and 10 CFR 61.56. These requirements are intended to ensure that future solid waste packages from all plants will be acceptable for burial at storage facilities and will maintain their long-term integrity.

Additional generic requirements that have evolved governing radioactive waste management systems are as follows. NUREG-0737, "Clarification of TMI Action Plan Requirements," Item II.F.1, issued in November 1980, provided new generic requirements regarding additional monitoring and sampling, and analysis of post-accident releases of radioactive materials. These requirements were added because of weaknesses noted at Three Mile Island, Unit 2 (TMI-2) in this area after the TMI-2 accident.

In the early 1980s, uncertainty arose with regard to future availability of low-level waste disposal capacity at the licensed burial sites. This concern resulted in the issuance of Generic Letter (GL) 81-38, "Storage of Low-Level Radioactive Wastes at Power Reactor Sites," dated November 10, 1981, which provided generic guidance to be used by licensees in the design, construction, and operation of such onsite storage facilities. Licensees continue to use this guidance to ensure proper storage of low-level waste at nuclear power plants.

11.4 Conclusions

The staff review at the time of the initial licensing of a facility determined that the information provided by the applicant was sufficient for the staff to conclude that the radioactive waste management system met the staff acceptance criteria and the intent of all applicable regulations. The staff has and will continue to obtain new information related to this subject area through a variety of sources such as updates to the FSAR, research reports, operating plant events and routine plant inspections. The staff reviews this information and, in the past, has required licensees to take actions to upgrade the plant to provide continuing assurance of adequate protection of the public health and safety. In addition, the staff will continue to review new information in this subject area and if the staff determines that new or different requirements are needed, the staff has the capability within the existing regulatory process to require additional analyses or plant modifications, as necessary, to ensure the continued health and safety of the public. In conclusion, the Commission concludes that the current regulatory process has and will continue to provide reasonable assurance that the licensing bases of all currently operating plants are sufficient to assure that operation is not inimical to the public health and safety.

12 RADIATION PROTECTION

12.1 Scope

General standards are provided for the protection of the individual from radiation hazards associated with activities licensed by the NRC. In this chapter, the staff will discuss the different control strategies in place to limit the exposure of occupational workers and the general public to ionizing radiation.

12.2 Safety Issues and Regulatory Requirements

Commission regulations provide standards for the protection of licensees, their employees, and the general public against the radiation hazards arising out of the possession or use of special nuclear, source, or byproduct material under license issued by the NRC. Certain precautionary procedures and administrative controls are provided to ensure that the evaluation of these radiation hazards are adequate and that the resulting radiation doses are kept as low as is reasonably achievable (ALARA). Different limits and controls are provided for occupationally exposed individuals and members of the general public.

12.2.1 Occupational Exposures

10 CFR Part 20 prescribes dose limits that govern the exposure of personnel to radiation from sources external to the body. In addition, limits on the quantities of radioactive material taken into the body through inhalation or absorption are provided to control the doses to individual organs and tissues from internal sources.

10 CFR Part 19.12 prescribes that plant workers be informed of the radiation hazards to which they are subjected and be instructed in the purpose and function of radiation protection devices in use and controls that they must observe.

12.2.2 Exposures to the General Public

10 CFR Part 20 provides controls for radiation exposure to the general public by limiting the radiation levels that can exist in areas not controlled by the licensee and concentrations of radioactive material that may be discharged from the facility in gaseous and liquid form, and by regulating the transportation and disposal of radioactive wastes.

In addition to the limits of 10 CFR Part 20 that apply to all NRC licensees, design specifications and operating requirements are provided in Appendix I to 10 CFR Part 50 to ensure that each power reactor licensee operates its facility so that the quantities of radioactive materials released to the environment in gaseous and liquid effluents are maintained ALARA.

Also, NRC licensees are subject to regulations promulgated by other agencies. The Environmental Protection Agency (EPA) provides limitations on the dose to members of the public from facilities in the uranium fuel cycle (including those licensed by the NRC) in 40 CFR Part 190. The Department of Transportation (DOT) regulates the shipment of radioactive materials in Title 49 to the Code of Federal Regulations (49 CFR).

12.3 Evolution of Current Licensing Basis

The adoption of 10 CFR Part 20 in 1957 established the basic framework currently employed for the protection of licensee personnel and the public from exposure to radiation. Extensive changes to the dose limits and the permissible concentrations of radioactive material in air and water, contained in 10 CFR Part 20, were adopted in 1960. These dose limits and permissible concentrations were based on the latest scientific knowledge of the time on the biological effects of radiation exposure.

An assumption basic to the radiation protection methods used in 10 CFR Part 20 is that any exposure to ionizing radiation results in a proportional health risk and that there should be no radiation exposure without a commensurate benefit. From 1970 to 1975, the Commission undertook a series of rule changes to improve the framework in 10 CFR Part 20 for ensuring that reasonable efforts are made to keep exposures to radiation, and releases of radioactivity in effluents, ALARA and to specify design and operating requirements in Appendix I to 10 CFR Part 50 to restrict quantities of radioactive materials released in gaseous and liquid effluents from light-water reactors (LWRs).

The dose criteria specified in Appendix I to 10 CFR Part 50 correspond to continuous effluent releases that are a small fraction of the concentration limits in 10 CFR Part 20. Licensees were required to implement technical specifications to ensure plant operations within the Appendix I requirements.

In 1981, the Commission amended 10 CFR Part 20 to incorporate the EPA requirements in 40 CFR Part 190, "Environmental Radiation Protection Standards for Nuclear Power Operations"; 40 CFR Part 190 provides that LWRs be operated so that releases of radioactive material and the resulting radiation doses to the public are below specified limits. These dose limits are comparable to, and in some cases more restrictive than, the dose objectives and operating conditions contained in Appendix I to 10 CFR Part 50.

In 1983, 10 CFR 20.311, was adopted to establish administration procedures and recordkeeping requirements to support the licensing requirements for land disposal of radioactive wastes contained in 10 CFR Part 61. The waste manifests, specified in 10 CFR 20.311, document that radioactive wastes are properly classified, described, packaged, marked, and labeled, and are in proper condition for transportation according to the applicable DOT regulations.

During the licensing process, pursuant to 10 CFR 50.34, the applicant must submit a final safety analysis report (FSAR) that describes the facility, the kinds and quantities of radioactive materials expected to be produced, and the means for controlling and limiting radioactive effluents and radiation exposures within the limits of 10 CFR Part 20.

Additional design criteria are provided in Appendix A to 10 CFR Part 50 governing the radioactive exposure to plant operators under accident conditions, the exposure to radiation during fuel handling and storage, and the adequate monitoring of radioactive concentrations in plant effluents and radiation levels in plant environs during normal operations and postulated accidents.

Before granting a license, the staff reviews the FSAR against established acceptance criteria and concludes that the facility design, and the radiological controls proposed, are adequate, and that the facility can be operated within all applicable limits and radiation exposures will be maintained ALARA.

The NRC inspection program ensures that each licensee adequately implements the radiation protection controls described in the plant FSAR and incorporated in the plant technical specifications. The performance of each licensee's radiation protection program is inspected by the NRC resident inspectors on a weekly basis, by region-based specialists routinely, and by teams of specialists whenever deemed necessary. When deficiencies are noted, the licensee is required to modify its program or implement additional controls as corrective action. An example of this is the finding from the 1980-1981 Health Physics Appraisal Team inspections that efforts at maintaining occupational radiation exposure ALARA lacked licensee support. Subsequently, each licensee implemented additional programs to ensure occupational exposures are maintained ALARA.

Plant operating events also provide new information that may require changes to a plant's licensing bases. Two examples are the serious exposure of plant workers during a fuel transfer operation in 1978 and the radiation protection experiences during the 1979 accident at Three Mile Island. In both cases, each LWR licensee was required to reanalyze its plant design and make appropriate modifications to ensure adequate protection to workers during spent fuel transfer operations or during postulated accident situations. In addition, the TMI accident indicated a need throughout the industry to improve accident assessment and monitoring capabilities related to potential radioactive releases off site during an accident. Upgrades in radiation protection during fuel handling operations were made at all operating reactor licensees as a required response to Bulletin 78-08, "Radiation Levels From Fuel Element Transfer Tubes," dated June 12, 1978. Improvements in radiation protection programs to protect workers during a postulated plant accident were contained in NUREG-0660 and NUREG-0737, and the implementation at each plant was subsequently required by Commission order.

Advances in the scientific and technical knowledge of radiobiology and the risks associated with radiation exposure have been made since the current limits in 10 CFR Part 20 were adopted in 1960. The current recommendations of the International Commission on Radiological Protection (ICRP) provide a radiation protection framework that relates the risks of nonuniform irradiation to individual tissues and organs from internally deposited radionuclides to the risk of uniform irradiation of the total body. This method differs from the current standards in 10 CFR Part 20, which limit internal and external exposures separately.

The Commission staff has reviewed the ICRP recommendations and has concluded that the framework more firmly establishes health risk as the basis for radiation protection than was evident for the current standards. The Commission is currently engaged in a rulemaking proceeding to revise 10 CFR Part 20 to be consistent with these international recommendations and practices, even though the standards in the current 10 CFR Part 20, in concert with the ALARA programs implemented at LWRs, result in doses generally far below the limits specified in the proposed revision. Thus, revising the limits in 10 CFR Part 20 will have little impact on reactor licensees. For example, limiting the sum of the dose from internally deposited radioactivity and the dose from external sources will not be a significant impact for LWRs since engineering and administrative controls, already required, generally reduce the intake of radioactive material to insignificant levels. If approved by the Commission, all NRC licensees will be required to change their programs to implement the radiation protection strategy provided by the revised 10 CFR Part 20. As additional scientific findings become available, such as those recently published by the National Academy of Sciences in its fifth report on the biological effects of ionizing radiation (BEIR V), the Commission will consider their significance and make changes to its rules and regulations as appropriate.

12.4 Conclusions

The staff review at the time of the initial licensing of a facility determined that the information provided by the applicant was sufficient for the staff to conclude that the control strategies in place to limit the exposure of occupational workers and the general public to ionizing radiation met the staff acceptance criteria and the intent of all applicable regulations. The staff has and will continue to obtain new information related to this subject area through a variety of sources such as updates to the FSAR, research reports, operating plant events and routine plant inspections. The staff reviews this information and, in the past, has required licensees to take actions to upgrade the plant to provide continuing assurance of adequate protection of the public health and safety. In addition, the staff will continue to review new information in this subject area and if the staff determines that new or different requirements are needed, the staff has the capability within the existing regulatory process to require additional analyses or plant modifications, as necessary, to ensure the continued health and safety of the public. In conclusion, the Commission concludes that the current regulatory process has and will continue to provide reasonable assurance that the licensing bases of all currently operating plants are sufficient to assure that operation is not inimical to the public health and safety.

13 CONDUCT OF OPERATIONS

The following sections discuss a number of subject areas that generally affect the conduct of operations around operating facilities. These subject areas include discussions of the management, operations, and technical support organization; training programs; emergency planning; licensee's self-assessment capabilities; plant procedures; and physical security.

13.1 Management, Operations, and Technical Support Organizations

13.1.1 Scope

During the licensing of an operating plant, the staff reviews the licensee's management and support organizations. This particular review area is limited to ensuring that corporate management is involved with, informed about, and dedicated to the safe design, test, and operation of the plant; that there are sufficient technical resources available to ensure plant operational safety; and that the structure, functions, and responsibilities of the licensee's onsite organization are acceptably defined.

13.1.2 Safety Issues and Regulatory Requirements

Commission regulations require that a licensee must be technically qualified to operate the plant before a license can be granted and that provisions relating to organization and management be included in the administrative controls section of the plant technical specifications. The regulations also describe licensed operator requirements during operation of a facility. The TMI Action Plan (NUREG-0737) also describes specific requirements with respect to the responsibility of both the shift supervisor and shift technical advisor.

13.1.3 Evolution of Current Licensing Basis

In general the safety evaluation report or its supplements contain descriptions of the management, operations, and technical support organizations for each facility at the time the license was issued. A licensee's management, operations, and technical support organizations continually change during the term of the license. Changes to the management, operations, and technical support organization are monitored throughout the term of the license, and new criteria are applied, if applicable.

10 CFR Part 50.71(e) requires each licensee to periodically update the Final Safety Analysis Report (FSAR) for their facility. The FSAR contains information with respect to the management, operations, and technical support organizations. If changes are made to provisions relating to management and organization that are in the technical specifications, the Commission reviews and approves these changes. Thereby, the Commission is continually updated on the licensee's management and technical support organizations at each facility.

The management, operations, and technical support organizations at each facility are evaluated continually. The Commission, on a continuing basis, interacts with the licensee through the evaluation of reportable events, license changes, 10 CFR Part 50.59 changes, the Commission resident inspector program, and special inspections. These activities provide insight into the capability of the technical support organization for the facility. The integration of all these interfaces with the licensee provides a continual evaluation of the management, operations, and technical support organization at each facility.

13.1.4 Conclusions

The staff review at the time of the initial licensing of a facility determined that the information provided by the applicant was sufficient for the staff to conclude that the applicant's technical qualifications to manage and support plant operations met the staff acceptance criteria and the intent of all applicable regulations. The staff has and will continue to obtain new information related to this subject area through a variety of sources such as updates to the FSAR, research reports, operating plant events and routine plant inspections. The staff reviews this information and, in the past, has required licensees to take actions to upgrade the plant to provide continuing assurance of adequate protection of the public health and safety. Reviews and approval of technical specification changes, interactions with the staff, and the inspection program provide the Commission with a continuing evaluation of the licensee's management, operations, and techncial support organizations. In addition, the staff will continue to review new information in this subject area and if the staff determines that new or differet requirements are needed, the staff has the capability within the existing regulatory process to require additional analyses of plant modifications, as necessary, to ensure the continue health and safety of the public. In conclusion, the Commission concludes that the current regulatory process has and will continue to provide reasonable assurance that the licensing bases of all currently operating plants are sufficient to assure that operation is not inimical to the public health and safety.

13.2 Training

13.2.1 Scope

This section describes information relating to the operational training and licensed operator requalification programs of the plant. The purpose of these programs is to provide assurance that the licensee will adequately train a staff to safely operate the plant and, thereby, protect the public health and safety.

13.2.2 Safety Issues and Regulatory Requirements

Commission regulations require that licensees provide training and instruction to individuals who manipulate the controls of a facility or direct any licensed activity of a licensed individual and provide information concerning organizational structure, personnel qualifications, and related matters to ensure that proper administrative and managerial controls are in place to ensure safe operation.

13.2.3 Evolution of Current Licensing Basis

Following the accident at Three Mile Island Unit 2 (TMI-2), the NRC emphasized the need to upgrade training and qualifications of nuclear power plant personnel. In the "NRC Action Plan Developed as a Result of the TMI-2 Accident"

(NUREG-0660, July 1980), the NRC cited its ongoing study of accreditation of training as a possible means of upgrading training programs in the industry.

In the "Clarification of TMI Action Plan Requirements" (NUREG-0737, November 1980), the NRC cited interim procedures to improve training programs and to upgrade the qualifications of personnel prior to accreditation of the facility training programs. Since that time, the Institute of Nuclear Power Operations (INPO), with its associated National Academy for Nuclear Power Operations (Academy), has developed a training accreditation program that the NRC has found to be an acceptable means of self-improvement of training.

On March 20, 1985, the Commission published its policy statement on training and qualification of nuclear power plant personnel allowing the industry a minimum of two years of accreditation activity without the introduction of new NRC training regulations. In the policy statement, the Commission further endorsed the training accreditation program managed by INPO, as it encompasses the elements of effective performance-based training and provides the basis to ensure that personnel have qualifications commensurate with the performance requirements of their jobs, and recognized the accreditation of 10 utility training programs.

On November 18, 1988, the Commission published a revised policy statement that reflected the minor modifications made by the Academy to its accreditation program and the NRC staff to the methods by which it monitors the industry training programs. Specifically, the amendments of the revised policy statement are: (1) recognition of the establishment of an eleventh accredited training program; (2) NRC staff will monitor the industry training programs and training program results by conducting post-accreditation reviews; and (3) NRC will conduct inspections, as deemed necessary, and take appropriate enforcement action in accordance with the Commission's enforcement policy in 10 CFR Part 2, Appendix C, when regulatory requirements are not met. However, the Commission's policy has been successfully challenged [Public Citizens v. U.S. NRC No. 89-1017 - D.C. Circuit, April 17, 1990]. The Commission has this matter under consideration.

To ensure that the nuclear industry's training program improvements are effective, the NRC monitors the accreditation process and its results by attending and observing Accreditation Board meetings, observing training accreditation team visits, conducting operator licensing and requalification exams, and conducting performance-oriented training inspections to assess the level of knowledge of plant personnel.

13.2.4 Conclusions

The staff review at the time of the initial licensing of a facility determined that the information provided by the applicant was sufficient for the staff to conclude that the applicant's training of licensed operators met the staff acceptance criteria and the intent of all applicable regulations. The staff has and will continue to obtain new information related to this subject area through a variety of sources such as updates to the FSAR, research reports, operating plant events and routine plant inspections. The staff reviews this information and, in the past, has required licensees to take actions to upgrade the plant to provide continuing assurance of adequate protection of the public health and safety. In particular, 10 CFR 50.34 and 10 CFR Part 55 establish the requirements for the development and implementation of training and requalification programs for facility personnel. The Commission has determined that the INPO training accreditation program with its periodic re-evaluation requirements is an adequate means of self-improvement training. In addition, the staff will continue to review new information in this subject area and if the staff determines that new or different requirements are needed, the staff has the capability within the existing regulatory process to require additional analyses of plant modifications, as necessary, to ensure the continue health and safety of the public. In conclusion, the Commission concludes that the current regulatory process has and will continue to provide reasonable assurance that the licensing bases of all currently operating plants are sufficient to assure that operation is not inimical to the public health and safety.

13.3 Emergency Planning

13.3.1 Scope

This section discusses the requirement that reactor licensees develop and implement emergency plans to ensure the continued protection of the public health and safety in the event of a radiological accident.

13.3.2 Safety Issues and Regulatory Requirements

Prior to the issuance of a full power operating license, the emergency planning regulations require a finding that there is reasonable assurance that adequate protective measures can and will be taken in the event of a radiological emergency. The regulations were adopted as an added conservatism to the defense-in-depth philosophy. They differ in character from most of the NRC's siting and engineering design requirements which are directed at achieving or maintaining a minimum level of public safety protection.

13.3.3 Evolution of Current Licensing Basis

13.3.3.1 Onsite Emergency Planning

In June 1979, NRC began a formal consideration of the role of emergency planning for ensuring the continued protection of the public health and safety in areas around nuclear power plant facilities. A final rule, effective November 3, 1980, was published in the Federal Register on August 19, 1980 (45 FR 55402). It provides that an initial operating license will not be granted unless NRC can make a favorable finding that the integration of onsite and offsite emergency planning provides reasonable assurance that adequate protective measures can and will be taken in the event of a radiological emergency. NRC will base its finding on a review of Federal Emergency Management Agency (FEMA) findings and determinations as to whether State and local emergency plans are adequate and capable of being implemented, and on the NRC assessment as to whether the applicant's onsite emergency plans are adequate and capable of being implemented. In the case of an operating reactor, if it is determined that there are such deficiencies that a favorable NRC finding is not warranted, and if the deficiencies are not corrected within four months of that determination, the Commission will determine expeditiously whether the reactor should shut down or whether some other enforcement action is appropriate. In any case, where the Commission believes that the public health, safety, or interest so requires, the plant will be required to shut down immediately. Licensees, however, will have an opportunity to demonstrate to the satisfaction of the Commission, for example, that deficiencies in emergency plans are not significant for the plant in question, that adequate interim compensating actions have been or will be taken promptly, or that there are compelling reasons to permit plant operation.

The 1980 rule required that emergency planning considerations be extended to emergency planning zones and that these consist of an area of about 10 miles in radius for exposure to the radioactive plume that might result from an accident in a nuclear power reactor and an area of about 50 miles in radius for food that might become contaminated. Additionally, the final rule sets forth 16 emergency planning standards that must be met by onsite, State, and local emergency plans within the emergency planning zones.

13.3.3.2 Offsite Emergency Planning

Section 109 of the NRC FY 1980 Authorization Bill (PL 96-295) required that NRC consult with the Director of the Federal Emergency Management Agency (FEMA) on the status of State radiological emergency response plans with respect to the issuance of an operating license for a reactor facility.

In response, FEMA issued a rule concerning review and approval of State radiological emergency plans and preparedness (44 CFR Part 350, September 28, 1983). This rule established policy and procedures for review and approval by FEMA of State emergency plans and preparedness for coping with the offsite effects of radiological emergencies that might occur at nuclear power facilities. The rule sets out criteria that are used by FEMA in reviewing, assessing, and evaluating the plans and preparedness; it specifies how and where a State may submit plans; and it describes certain of the processes by which FEMA makes findings and determinations as to the adequacy of State plans and the capability of State and local governments to implement these plans and preparedness measures. Such findings and determinations are to be submitted to the Governors of affected States and to NRC for use in its licensing proceedings.

13.3.3.3 Current Program

As experience was gained in the implementation of the revised onsite and offsite emergency plans by both the licensees and the State and local governments, revisions to the regulations were deemed appropriate. For example, the 1980 regulations required that the licensees and State and local governments within the 10-mile plume exposure pathway emergency planning zone conduct an annual, full-participation exercise. After considering the experience gained by all of the participants in these annual exercises, the Commission proposed, and then adopted in 1984 a change to a biennial full-participation exercise. The revised rule continued to require an annual onsite exercise for licensees, required State and local governments to participate every 2 years with a provision for remedial exercises to ensure that any deficiencies are corrected, and provided the opportunity for State and local government participation in the annual licensee exercise, if desired. The rationale behind the change was that (1) experience in observing and evaluating over 150 exercises had shown a disproportionate amount of Federal, State, and local government resources were being expended to conduct and evaluate the annual exercises, (2) State and local governments respond to a variety of actual emergencies on a continuing basis, thus exercising their emergency preparedness capabilities, and (3) the flexibility provided for in a biennial frequency would provide an incentive for State and local governments to perform in a satisfactory manner in order to avoid conducting remedial exercises.

In order to ensure that emergency preparedness around licensed nuclear facilities continues to reflect current conditions and circumstances, licensees are permitted to make changes to emergency plans without NRC approval if those changes do not decrease the effectiveness of these plans, and the plans, as changed, continue to meet the standards of 10 CFR 50.47(b) and the requirements of 10 CFR Part 50, Appendix E. Changes made without approval must be reported within 30 days after the changes are made. Proposed changes that decrease the effectiveness of the approved plans may not be implemented without application to and approval by the NRC. This requirement is found in 10 CFR 50.54(q) of the regulations.

As required by 10 CFR 50.47(b)(8), licensees are to provide and maintain adequate emergency facilities and equipment. To satisfy this requirement, licensees must inspect and perform operability checks of emergency equipment and instruments at frequent intervals throughout the year. In addition, the NRC performs an annual inspection of the licensee's program and equipment to ensure that essential emergency facilities, equipment, instrumentation, and supplies are being maintained in a state of operational readiness. Based on the above discussion, the staff does not believe that equipment used in assuring the effectiveness of the emergency preparedness program needs to be evaluated as part of the plant assessment of aging degradation required by the new Part 54.

13.3.4 Conclusions

The staff review at the time of the initial licensing of a facility determined that the information provided by the applicant was sufficient for the staff to conclude that the facility's emergency plans met the staff acceptance criteria and the intent of all applicable regulations. The staff has and will continue to obtain new information related to this subject area through a variety of sources such as updates to the FSAR, research reports, operating plant events and routine plant inspections. The staff reviews this information and, in the past, has required licensees to take actions to upgrade the plant to provide continuing assurance of adequate protection of the public health and safety. In addition, the staff will continue to review new information in this subject area and if the staff determines that new or differet requirements are needed, the staff has the capability within the existing regulatory process to require additional analyses of plant modifications, as necessary, to ensure the continue health and safety of the public. In conclusion, the Commission concludes that the current regulatory process has and will continue to provide reasonable assurance that the licensing bases of all currently operating plants are sufficient to assure that operation is not inimical to the public health and safety.

13.4 Review and Audit

13.4.1 Scope

This section discusses the licensee's operational review program. The purpose of this program is to implement the licensee's responsibility-related proposed changes, test evaluations of unplanned events, and provisions for the evaluation of plant operations.

13.4.2 Safety Issues and Regulatory Requirements

The Commission regulations require that one of the considerations in granting a license is the technical qualifications of the applicant. NUREG-0737, "Clarification of TMI Task Action Plan Requirements," describes the requirements for an Independent Safety Engineering Group (ISEG) for post-TMI licensed plants. In addition, the regulations require that certain provisions relating to administrative controls be incorporated into the administrative controls section of the plant-specific technical specifications.

13.4.3 Evaluation of Current Licensing Basis

In general, the safety evaluation report for each facility contains a description of the operational review program at the time the facility was licensed.

A licensee's operational review program is often revised during the term of the license. Changes to the program are monitored by the Commission throughout the term of the license, and new criteria are applied if applicable.

10 CFR Part 50.71(e) requires each licensee to periodically update the Final Safety Analysis Report (FSAR) for their facility. The FSAR contains a description of the facility operational review program. In addition, the Commission reviews and approves any changes to the operational review program that are in the facility technical specifications. Thereby, the Commission is periodically updated on the current operational review program at each facility.

The Commission's inspection program provides for periodic evaluation of the facility operational review program. Inspection Procedure 40500, "Evaluation of Licensee's Self-Assessment Capability," and Inspection Procedure 88005, "Management Organization and Controls," provide for the periodic inspection and evaluation of the facility operational review program.

13.4.4 Conclusions

The staff review at the time of the initial licensing of a facility determined that the information provided by the applicant was sufficient for the staff to conclude that the applicants operational review program met the staff acceptance criteria and the intent of all applicable regulations. The staff has and will continue to obtain new information related to this subject area through a variety of sources such as updates to the FSAR, research reports, operating plant events and routine plant inspections. The staff reviews this information and, in the past, has required licensees to take actions to upgrade the plant to provide continuing assurance of adequate protection of the public health and safety. In addition, the staff will continue to review new information in this subject area and if the staff determines that new or differet requirements are needed, the staff has the capability within the existing regulatory process to require additional analyses of plant modifications, as necessary, to ensure the continue health and safety of the public. In conclusion, the Commission concludes that the current regulatory process has and will continue to provide reasonable assurance that the licensing bases of all currently operating plants are sufficient to assure that operation is not inimical to the public health and safety.

13.5 Plant Procedures

This section discusses two general categories of procedures: administrative procedures and operating and maintenance procedures. Administrative procedures include (1) those that provide the administrative controls with respect to procedures, and (2) those that define and provide controls for operational activities of the plant staff. Operating and maintenance procedures are used by the operating organization (plant staff) to ensure that routine operating, off-normal, emergency, and maintenance activities are conducted in a safe manner.

13.5.1 Administrative Procedures

13.5.1.1 Safety Issues and Regulatory Requirements

The Commission regulations require that one of the considerations in granting a license is the technical qualifications of the applicant to engage in the activities of the license and require that the licensee designate individuals to be responsible for directing the activities of licensed operators. The regulations also require provisions relating to administrative controls in the Administrative Controls Section of the Technical Specifications. Further, NUREG-0737, "Clarification of TMI Task Action Plan Requirements," describes certain requirements with respect to administrative procedures requirements.

3.5.1.2 Evolution of Current Licensing Basis

The safety evaluation report (SER) and its supplements describe the administrative controls program at the time of licensing of the facility. The administrative controls relate to, in part, procedures and programs and to designating individuals to be responsible for directing the activities of licensed operators.

A licensee's procedures and program for the control of procedures may change during the term of the license. Changes to these procedures and programs are monitored during the term of the license and new criteria applied if applicable.

10 CFR Part 50.71(e) requires each licensee to periodically update the Final Safety Analysis Report (FSAR) for their facility. The FSAR contains the administrative controls program for procedures. Changes to the procedures control program that are included in the technical specifications are reviewed and approved by the Commission. Thereby, the Commission is continually aware of the administrative controls program.

The licensee's administrative controls program at each facility is periodically evaluated through the Commission's inspection program. In particular, Inspection Procedure 71707, "Operational Safety Verification," and Inspection Procedure 88005, "Management Organization and Controls," provide for a continual review of the licensee's administrative procedures control program.

13.5.1.3 Conclusions

The staff review at the time of the initial licensing of a facility determined that the information provided by the applicant was sufficient for the staff to conclude that the applicant's administrative controls program met the staff acceptance criteria and the intent of all applicable regulations. The staff has and will continue to obtain new information related to this subject area through a variety of sources such as updates to the FSAR, research reports, operating plant events and routine plant inspections. The staff reviews this information and, in the past, has required licensees to take actions to upgrade the plant to provide continuing assurance of adequate protection of the public health and safety. In addition, the staff will continue to review new information in this subject area and if the staff determines that new or differet requirements are needed, the staff has the capability within the existing regulatory process to require additional analyses of plant modifications, as necessary, to ensure the continue health and safety of the public. In conclusion, the Commission concludes that the current regulatory process has and will continue to provide reasonable assurance that the licensing bases of all currently operating plants are sufficient to assure that operation is not inimical to the public health and safety.

13.5.2 Operating and Maintenance Procedures

13.5.2.1 Safety Issues and Regulatory Requirements

The regulations applicable to administrative procedures require the determination that the licensee is technically qualified to engage in licensing activities and that the licensee designate individuals to be responsible for directing the licensed activities of licensed operators.

Commission regulations also govern operating procedures used by licensed operators in the control room and other operating procedures and maintenance procedures. Additionally, the TMI Action Plan (NUREG-0660 and NUREG-0737) requires licensees to upgrade their Emergency Operating Procedures (EOPs).

Commission regulations also require that activities affecting quality be prescribed by and accomplished in accordance with documented instructions, procedures, and drawings.

13.5.2.2 Evolution of Current Licensing Basis

Requirements for the commercial nuclear power industry to improve the quality and usability of plant procedures were established as a result of the accident at Three Mile Island (TMI). Following TMI, the NRC Office of Nuclear Reactor Regulation developed the TMI Action Plan (NUREG-0660 and NUREG-0737) which required licensees of operating reactors to reanalyze transients and accidents and to upgrade EOPs (Item I.C.1). NUREG-0660 (Item I.C.9) committed the NRC to develop a long-term plan for the overall improvement of nuclear power plant procedures.

Requirements for EOPs were further defined in Generic Letter 82-33. Generic Letter 82-33 transmitted Supplement 1 to NUREG-0737, "Requirements for Emergency

Response Capability," and directed each licensee to submit to the NRC a Procedures Generation Package (PGP) from which licensees were to develop function or symptom-based EOPs. This document also indicated that the NRC staff would audit EOPs on a selective basis.

Early reviews of EOP programs identified potential concerns with their implementation. In response to these findings, the NRC staff conducted inspections to monitor the industry's procedure upgrade programs. Initial inspections revealed a number of problems, and Information Notice 86-64 was issued in August 1986 to alert licensees to these problems. Subsequent inspections revealed similar results and Information Notice 86-64, Supplement 1, was issued on April 20, 1987, to describe further problems with EOPs and PGPs and to inform the industry that the inspection effort would be intensified. NRC Temporary Instruction 2515/92 was issued in April 1988 and defines the objectives of the EOP inspection. The inspection effort now extends to all operating reactors in the United States and has two objectives: (1) to assess the adequacy of the EOPs themselves, and (2) to establish that the supporting programs and documents are sufficient to ensure the integrity and continued adequacy of the EOPs.

13.5.2.3 Conclusions

The staff review at the time of the initial licensing of a facility determined that the information provided by the applicant was sufficient for the staff to conclude that the applicant's procedure revision process met the staff acceptance criteria and the intent of all applicable regulations. The staff has and will continue to obtain new information related to this subject area through a variety of sources such as updates to the FSAR, research reports, operating plant events and routine plant inspections. The staff reviews this information and, in the past, has required licensees to take actions to upgrade the plant to provide continuing assurance of adequate protection of the public health and safety. In addition, the staff will continue to review new information in this subject area and if the staff determines that new or different requirements are needed, the staff has the capability within the existing regulatory process to require additional analyses of plant modifications, as necessary, to ensure the continue health and safety of the public. In conclusion, the Commission concludes that the current regulatory process has and will continue to provide reasonable assurance that the licensing bases of all currently operating plants are sufficient to assure that operation is not inimical to the public health and safety.

13.6 Physical Security

13.6.1 Scope

This section discusses the evolution of the basis of the reactor security program. The licensee's security program consists of the following three plans: security, security contingency, and guard training and qualification. These three plans provide the physical protection envelope that provides the assurances that the operation of these plants does not constitute an unreasonable risk to the public health and safety.

13.6.2 Safety Issues and Regulatory Requirements

The Commission regulations require licensees to establish and maintain a physical protection system and security organization that provides high assurance against radiological sabotage.

13.6.3 Evolution of Current Licensing Basis

The purpose of nuclear power reactor security requirements is to protect against the design basis threat of radiological sabotage. The design basis threat is generally considered to be the worst-case scenario of attack by several well-trained and dedicated individuals and an individual inside the facility. In 1977, specific requirements of physical protection of licensed nuclear facilities against radiological sabotage were set forth by the NRC in 10 CFR 73.55. In publishing this rule, the Commission stated the following: "The level of protection specified in Part 73.55 is adequate and prudent at this time. The kind and degree of threats will continue to be reviewed by the Commission. Should such reviews show change that would dictate different levels of protection, the Commission would consider changes to meet the changed conditions (42 FR 10836)."

The Commission has since made a number of changes to the requirement to maintain or increase the level of assurance. In 1978, the Commission issued requirements for a safeguards contingency plan and guard training and qualification plans to be prepared and noted in a facility's license conditions. Subsequent changes in Part 73 have required the reporting of physical security events, the protection of unclassified safeguards information, and the "Miscellaneous Amendments." Those amendments include a refined vital area access policy, authority to suspend safeguards during safety emergencies, protection of certain safeguards equipment, and upgrades to key and lock controls. Most recently the regulations have been revised to require that any individual in need of unescorted access at a facility submit to a Federal Bureau of Investigation fingerprint check and chemical testing to determine that they are fit-for-duty.

These changes were made to ensure that the level of protection remains adequate considering all new information and potential threats. In 1989 the Commission requested licensees to include in their safeguards contingency plan procedures for short-term actions to protect against attempted radiological sabotage involving a land vehicle bomb if such a threat were to materialize.

The NRC has conducted Regulatory Effectiveness Reviews (RERs) since 1982 to ensure that safeguards required by NRC's regulations, as implemented by licensees, provide the intended level of protection without compromising safety of operations. The RER teams use NRC security personnel and members of the U.S. Army Special Forces to test plant security systems and personnel. Regional safeguards inspectors continue their routine unannounced and special inspections at all licensed facilities.

In addition to continued NRC review of industry-wide conditions, the status of physical security measures are reviewed at each individual plant in the Systematic Assessment of Licensee Performance (SALP) program. Both headquarters and regional safeguards staff provide comments for the "Security" functional area.

Licensee-initiated changes to approved security plans (also contingency and guard training) may only be made by two methods. Changes made pursuant to 10 CFR 50.54(p) may be made without prior Commission approval if the changes do not decrease the safeguards effectiveness of the plan. The changes must be submitted to the Commission within two months and changes are reviewed by the staff. The second method for plan changes involves the amendment process as specified in 10 CFR 50.90 to include reviews by the staff and Federal Register notices soliciting public comment. These changes may involve measures that are not contained in 10 CFR 73.55(b) through (h), but provide the equivalent high assurance against radiological sabotage.

Age-related degradation of safeguards equipment is not a license renewal issue because it is an issue that is being currently experienced and managed. A number of the originally licensed sites have reached the life expectancy of certain types of security equipment. Because of the general performance objectives and requirements of 10 CFR 73.55(a) and the site-specific commitments contained in the individual plant security plans, normal inspection activities will force the replacement of degraded equipment or subject the licensee to enforcement action.

13.6.4 Conclusions

The staff review at the time of the initial licensing of a facility determined that the information provided by the applicant was sufficient for the staff to conclude that the physical security program met the staff acceptance criteria and the intent of all applicable regulations. The staff has and will continue to obtain new information related to this subject area through a variety of sources such as updates to the FSAR, research reports, operating plant events and routine plant inspections. The staff reviews this information and, in the past, has required licensees to take actions to upgrade the plant to provide continuing assurance of adequate protection of the public health and safety. In addition, the staff will continue to review new information in this subject area and if the staff determines that new or different requirements are needed, the staff has the capability within the existing regulatory process to require additional analyses of plant modifications, as necessary, to ensure the continue health and safety of the public. In conclusion, the Commission concludes that the current regulatory process has and will continue to provide reasonable assurance that the licensing bases of all currently operating plants are sufficient to assure that operation is not inimical to the public health and safety.

14 INITIAL TEST PROGRAM

14.1 Safety Issues and Regulatory Requirements

Commission regulations require, in part, that an applicant for a license to operate a production or utilization facility include the principal design criteria for the proposed facility in the safety analysis report (SAR). These regulations state that these principal design criteria are to establish the necessary design, fabrication, construction, testing, and performance requirements for systems, structures, and components important to safety, that is, systems, structures, and components that provide reasonable assurance that the facility can be operated without undue risk to the health and safety of the public.

These regulations also require that a test program be established to ensure that systems, structures, and components will perform satisfactorily in service. Since all functions designated in the general design criteria are important to safety, all systems, structures, and components required to perform these functions need to be tested to ensure that they will perform properly. These functions, as noted throughout the specific general design criteria, are those necessary to ensure that specified design conditions of the facility are not exceeded during any condition of normal operation, including anticipated operational occurrences, or as a result of postulated accident conditions.

14.2 Evolution of Current Licensing Basis

The NRC safety evaluation report (SER) and supplements describe and attest to the adequacy of the initial test program for each facility at the time the license was issued. Upon completion of the test program, the results are documented in a final test report subsequent to issuance of an operating license for each facility. The satisfactory completion of the test program provides assurance that the systems, structures, and components important to safety will perform as designed and that the facility can be operated without undue risk to the health and safety of the public. During the term of the initial license, Commission oversight, regulatory actions, and the implementation of technical specifications provide assurance that the plant continues to meet the current licensing basis. This is sufficient to conclude that the level of safety is also adequate for continued operation during any renewal period.

14.3 Conclusions

The staff review at the time of the initial licensing of a facility determined that the information provided by the applicant was sufficient for the staff to conclude that the licensee's test program met the staff acceptance criteria and the intent of all applicable regulations. The staff has and will continue to obtain new information related to this subject area through a variety of sources such as updates to the FSAR, research reports, operating plant events and routine plant inspections. The staff reviews this information and, in the past, has required licensees to take actions to upgrade the plant to provide continuing assurance of adequate protection of the public health and safety.

In addition, the staff will continue to review new information in this subject area and if the staff determines that new or different requirements are needed, the staff has the capability within the existing regulatory process to require additional analyses or plant modifications, as necessary, to ensure the continued health and safety of the public. In conclusion, the Commission concludes that the current regulatory process has and will continue to provide reasonable assurance that the licensing basis of all currently operating plants are sufficient to assure that operation is not inimical to the public health and safety.

15 ACCIDENT ANALYSES

15.1 Scope

This chapter addresses the analyses of the response of the plant to postulated accidents and to postulated malfunctions or failures of equipment. Such safety analyses provide a significant contribution to the selection of limiting conditions of operation, limiting safety system settings, and design specifications for components and systems from the standpoint of public health and safety. Also, the effects of anticipated accidents and postulated component failures are examined to determine their consequences and to evaluate the capability built into the plant to control or accommodate such failures and situations. The situations analyzed include anticipated operational occurrences (e.g., a loss of electrical load resulting from a line fault), off-design transients that include a small amount of fuel failures, and postulated accidents of low probability (e.g., the sudden loss of integrity of reactor coolant pressure boundary). The analyses include an assessment of the consequences of an assumed fission-product release.

15.2 Safety Issues and Regulatory Requirements

Commission regulations require, in part, that the reactor core and associated coolant, control, and protection systems shall be designed with appropriate margin to ensure that specified acceptable fuel design limits (SAFDLs) are not exceeded during any condition of normal operation, including the effects of anticipated operational occurrences; that the reactor coolant system and associated auxiliary, control, and protection systems shall be designed with sufficient margin to ensure that the design conditions of the reactor coolant pressure boundary are not exceeded during any condition of normal operation, including anticipated operational occurrences; and that redundant and reliable reactivity control systems are provided to ensure that under conditions of normal operation, including anticipated operational occurrences, SAFDLs are not exceeded.

15.3 Evolution of Current Licensing Basis

The licensing basis for transient and accident analyses has evolved as reactor events provide new information that is determined to provide improvement in the methods of evaluation. The process of evaluating operating experience and assessing plant data to determine the need for additional actions is a continuing one.

General staff guidance specifies that the transients and accidents analyzed in the plant safety analysis report ensure that a sufficiently broad spectrum of initiating events has been considered; ensure that initiating events of certain types and expected frequencies of occurrence be analyzed so that only the limiting cases in each group are quantitatively evaluated; and permit the consistent application of specific acceptance criteria for each postulated initiating event. In general, each initiating event is assigned to one of three frequency groups: incidents of moderate frequency, infrequent incidents, or limiting faults. The quantitative evaluation of each initiating event in each of the three frequency groups establishes the limiting conditions of operation for the required safety systems and the limiting parameters are routinely placed in the plant technical specifications to ensure that the plant is operated within its established design envelope.

The evolution of the current licensing basis regarding the performance of the emergency core cooling system following a postulated loss-of-coolant accident is discussed in Section 6.4.2 of this report.

The 1979 accident at Three Mile Island, Unit 2 (TMI-2) raised several issues (TMI action items) that affect the management of plant transients and accidents. Items II.E.1.1 and II.E.1.2 of NUREG-0737 required upgrades to the auxiliary feedwater system in all pressurized-water-reactor (PWR) plants to improve its reliability. The specific improvements include changes in design regarding system initiation and flow indication. Item II.K.3.5 provides guidelines on automatic trip of the reactor coolant pump during a postulated loss-of-coolant accident. Item II.K.3.44 requires an evaluation of anticipated transient with single failure to verify no significant fuel failure.

Within the framework of the Systematic Evaluation Program (SEP), the staff evaluated the need for all plants to adopt the reactor coolant system specific activity limits found with the Standard Technical Specifications (SEP Lessons Learned Issue 6.3). The coolant activity levels have a proportionate effect on those accidents involving primary coolant release (without core damage) to the environment. Implementation of Standard Technical Specifications (STS) limits are usually adequate to alleviate the concerns regarding resultant offsite doses.

The staff examined the plant's technical specifications to determine the degree of compliance with the appropriate STS. An evaluation was performed to determine the adequacy of the existing plant technical specification limits in restricting offsite dose. The review covered those accidents whose primary dose contribution is from reactor coolant leakage to the atmosphere (e.g., main steam line break outside of containment, steam generator tube rupture, and small line breaks outside containment).

With respect to boiling-water reactors (BWRs), the staff reviewed Generic Safety Issue (GSI) 74. On the basis of this review, the staff determined that GSI 74 had low safety significance and that no further staff review was warranted. Additionally, as a result of improvements in nuclear fuel performance, steam generator performance for PWRs, and chemistry control for both PWRs and BWRs, the potential safety significance of this issue has been further reduced. Therefore, the staff concluded that this SEP lessons learned issue have been adequately addressed by other regulatory initiatives. The staff no longer believes that this is an issue for any generating plant.

15.4 Conclusions

The staff review at the time of the initial licensing of a facility determined that the information provided by the applicant was sufficient for the staff to conclude that transient and accident analyses conducted by the licensee met the staff acceptance criteria and the intent of all applicable regulations. The staff has and will continue to obtain new information related to this subject area through a variety of sources such as updates to the FSAR, research reports, operating plant events and routine plant inspections. The staff reviews this information and, in the past, has required licensees to take actions to upgrade the plant to provide continuing assurance of adequate protection of the public health and safety. Before each plant refueling, the transient and accident analyses are reviewed by each licensee to verify the changes resulting from new core don not result in an unreviewed safety question. If an unreviewed safety question arises, or if any technical specifications require modification, staff review and approval is required before plant restart. In addition, the staff will continue to review new information in this subject area and if the staff determines that new or different requirements are needed, the staff has the capability within the existing regulatory process to require additional analyses or plant modifications, as necessary, to ensure the continued health and safety of the public. In conclusion, the Commission concludes that the current regulatory process has and will continue to provide reasonable assurance that the licensing bases of all currently operating plants are sufficient to assure that operation is not inimical to the public health and safety.

16 TECHNICAL SPECIFICATIONS

16.1 Scope

Each applicant for an operating license is required to submit proposed technical specifications and their bases for the facility as a chapter in the final safety analysis report (FSAR). They should be consistent with the content and format of the Standard Technical Specifications available from the Commission for the appropriate nuclear steam supply system (NSSS) vendor. After review and needed modification by the NRC staff, these technical specifications are issued by the Commission as Appendix A to the operating license.

The equipment included in the technical specifications is a broad spectrum of structures and electrical and mechanical systems and components taken from the safety analyses of the FSAR or updated safety analysis report (USAR). It includes such structures as the reactor vessel and containment, such systems such as the emergency core cooling system and reactor protection system, and such components as circuit breakers, valves, and pumps in these systems.

16.2 Safety Issues and Regulatory Requirements

Commission regulations require that each license issued by the NRC authorizing operation of a utilization facility shall include technical specifications. These regulations also describe the required contents of the technical specifications. Safety limits, settings for automatic protective devices, and limiting conditions for operation are required to be included in these technical specifications. Surveillances are also required to ensure that the necessary quality of systems and components is maintained, that important parameters are maintained within specified limits, and that the limiting conditions for operation are satisfied. Compensatory actions, which may include shutting down the reactor, are required when it is found that these conditions are not met.

The technical specifications are derived from the analyses in the final (or updated) safety analysis report. They ensure that the plant will be operated so that the assumptions of these safety analyses remain valid. The assumptions include both initial conditions and availability of equipment.

16.3 Evolution of Current Licensing Basis

Technical specifications are required by the Atomic Energy Act of 1954 (Section 182), which states, in part, that

> The applicant shall state such technical specifications, including information of the amount, kind, and source of special nuclear material required, the place of the use, the specific characteristics of the facility, and such other information as the Commission may, by rule or regulation, deem necessary in order to enable it to find that the utilization or production of special nuclear material will .... provide adequate protection to the health and safety of the public. Such technical specifications shall be a part of any license issued.

In 1968, the Commission issue 10 CFR 50.36 to specify the content of the technical specifications. From this time until 1973, each plant's technical specifications were unique but similar. In 1973, the concept of Standard Technical Specifications was introduced in an attempt to make technical specifications of different plants more consistent. When a plant is ready to be licensed, it uses the applicable Standard Technical Specifications as a starting point.

Licensees typically request changes to the technical specifications in accordance with existing regulations as the plant is operated throughout its life to reflect modifications to the design and different methods of operation. When such changes are requested, the NRC must review and approve the requested changes before they can be implemented. In addition, the NRC also requires changes to the technical specifications as new safety and licensing issues arise. For example, changes were required to plant technical specifications as a result of the Three Mile Island, Unit 2 accident. The resolution of other important issues, such as the potential for overpressurization of reactor vessels at low temperatures and isolation of low-pressure systems from high-pressure systems, have led to additions to technical specifications to ensure that plant operation is in conformance with the resolution of these problems.

As discussed above, the surveillances required by the technical specifications ensure that the plant is operated so that the technical specifications requirements are met. Technical specifications requirements on equipment are primarily a check on operability of the equipment. Degradation (as, for example, from aging) is in most cases not specifically required to be measured, although the American Society of American Engineers Boiler and Pressure Vessel Code (ASME Code) (which is incorporated in most technical specifications) requires a limited amount of trending of performance for pumps and valves. However, the surveillances are generally done frequently enough so that degradation is not expected to occur to the extent that operability is affected between surveillances.

16.4 Conclusions

The staff review at the time of the initial licensing of a facility determined that the information provided by the applicant was sufficient for the staff to conclude that the plant's technical specifications met the staff acceptance criteria and the intent of all applicable regulations. The staff has and will continue to obtain new information related to this subject area through a variety of sources such as updates to the FSAR, research reports, operating plant events and routine plant inspections. The staff reviews this information and, in the past, has required licensees to take actions to upgrade the plant to provide continuing assurance of adequate protection of the public health and safety. In addition, the staff will continue to review new information in this subject area and if the staff determines that new or different requirements are needed, the staff has the capability within the existing regulatory process to require additional analyses of plant modifications, as necessary, to ensure the continue health and safety of the public. In conclusion, the Commission concludes that the current regulatory process has and will continue to provide reasonable assurance that the licensing bases of all currently operating plants are sufficient to assure that operation is not inimical to the public health and safety.

17 QUALITY ASSURANCE

17.1 Scope

The quality assurance (QA) program of licensees applies to systems, structures, and components that prevent or mitigate the consequences of postulated accidents that could cause undue risk to the health and safety of the public. The QA program of each licensee is reviewed by the NRC to ensure that it meets the requirements of Appendix B to 10 CFR Part 50, and the NRC performs inspections to determine whether the program is being implemented effectively.

17.2 Safety Issues and Regulatory Requirements

Commission regulations require that licensees establish and maintain a QA program for the design, construction, and operation of systems, structures, and components that prevent or mitigate the consequences of postulated accidents that could cause undue risk to the health and safety of the public.

17.3 Evolution of Current Licensing Basis

In publishing a proposed rule in 1970 to add Appendix B to 10 CFR Part 50, the Commission stated that its purpose was to establish QA requirements for systems, structures, and components to prevent or mitigate the consequences of postulated accidents that could cause undue risk to the health and safety of the public. Further, the Commission stated that the requirements of Appendix B would apply to all activities during design, construction, and operation of such systems, structures, and components and that the criteria of Appendix B would be used for guidance in evaluating the adequacy of the QA programs in use by holders of both construction permits and operating licenses. In essence, Appendix B establishes the minimum acceptable QA requirements for providing reasonable assurance that (1) applicable regulatory requirements and the design basis for systems, structures, and components, as specified in the license application, are correctly translated into specifications, drawings, procedures, and instructions and (2) subsequent activities, such as construction, operation, testing, refueling, repair, maintenance, modification, and decommissioning, are conducted and verified in accordance with appropriate procedures and instructions.

In the early 1980s, the Commission identified a concern with plant-specific implementation and modification of NRC-approved QA programs. The Commission noted that changes being made in previously approved QA programs could diminish their effectiveness and result in unacceptable QA programs at some licensed facilities. In publishing a final rule addressing this concern, the Commission stated that an NRC-approved QA program becomes a principal inspection and enforcement tool in ensuring that a licensee is in compliance with QA requirements for protecting the public health and safety. In addition, the final rule [10 CFR 50.54(a) and 10 CFR 50.55(f)] established a procedure requiring review and approval by the Commission before implementing any change to a previously approved QA program that would reduce its effectiveness.

The final rule also required that each licensee submit a current description of its QA program and thereafter submit any revisions annually for NRC review and re-approval. Through these requirements, the Commission established an acceptable baseline for a QA program at each plant against which future changes to the program would be judged, and ensured that future changes would be available. This rule change created a regulatory process by which the Commission ensures that an acceptable QA program will remain in place at a licensed facility throughout the life of the license and that changes to that program would be routinely reviewed and evaluated to ensure that the program would continue to satisfy the regulatory requirements of Appendix B to 10 CFR Part 50. Toward ensuring this end, the NRC routinely inspects the implementation of QA programs. Safety-related activities undertaken by licensees to obtain a renewed license are also subject to the requirements of Appendix B.

17.4 Conclusions

The staff review at the time of the initial licensing of a facility determined that the information provided by the applicant was sufficient for the staff to conclude that QA program met the staff acceptance criteria and the intent of all applicable regulations. The staff has and will continue to obtain new information related to this subject area through a variety of sources such as updates to the FSAR, research reports, operating plant events and routine plant inspections. The staff reviews this information and, in the past, has required licensees to take actions to upgrade the plant to provide continuing assurance of adequate protection of the public health and safety. Under 10 CFR 50.54(a) and 10 CFR 50.55(e), any subsequent change to relax or reduce the previous commitments of the QA program of a licensee must receive NRC approval before the licensee can implement the change. In addition, the staff will continue to review new information in this subject area and if the staff determines that new or different requirements are needed, the staff has the capability within the existing regulatory process to require additional analyses or plant modifications, as necessary, to ensure the continued health and safety of the public. In conclusion, the Commission concludes that the current regulatory process has and will continue to provide reasonable assurance that the licensing bases of all currently operating plants are sufficient to assure that operation is not inimical to the public health and safety.

18 HUMAN FACTORS ENGINEERING

18.1 Scope

This chapter describes the regulatory requirements in the human factors area. Human factors engineering has played a significant role in control room design and in the technical areas of safety parameter display.

18.2 Control Room

18.2.1 Scope

Nuclear power plants have a control room from which employees can operate the plant safely under normal and accident conditions. Outside the control room, equipment has the design capability for prompt hot shutdown of the reactor, which includes the necessary instrumentation and controls to maintain the plant in a safe condition during hot shutdown, and with a potential capability for cold shutdown.

18.2.2 Safety Issues and Regulatory Requirements

The safety issue addressed is to confirm that the design of the plant's control room and remote shutdown capability facilitates the plant operator's ability to prevent accidents or cope with accidents if they do occur.

The basis for regulating the design of the plant's control room and remote shutdown capability is given in the enclosure to Generic Letter 82-33, "Supplement 1 to NUREG-0737--Requirements for Emergency Response Capability."

18.2.3 Evolution of Current Licensing Basis

Requirements for commercial nuclear power plants to review their control room design and correct deficiencies were established as a result of the Three Mile Island (TMI) accident. In May 1980, NUREG-0660, "TMI Action Plan Developed as a Result of the TMI-2 Accident," was issued. Item I.D.1, "Control Room Design Reviews," stated that "NRR will require that operating reactor licensees and applicants for operating licenses perform a detailed control room design review to identify and correct design deficiencies." The review was to be performed on a schedule consistent with the implementation of other requirements for enhancing operator effectiveness, including necessary retraining. In November 1980, the NRC published NUREG-0737, "Clarification of TMI Action Plan Items," which identified the requirements associated with detailed control room design reviews (DCRDRs). Guidance published as NUREG-0700, "Guidelines for Control Room Design Reviews" (1981), was also issued to the industry. In December 1982, "Supplement 1 to NUREG-0737--Requirements for Emergency Response Capability," was issued as Generic Letter 82-33. This document implemented existing requirements for plants to conduct a DCRDR and identify human engineering discrepancies and provided additional clarification. For some, but not all, plants, the NRC issued confirmatory orders, which required plants to submit schedules for completing a program plan and a summary report (including a proposed schedule for implementation) of their DCRDRs.

For operating plants, the staff has reviewed the program plans for conducting and implementing the DCRDR. These plants also have submitted a summary report of their completed review, which outlines proposed control room changes and implementation schedules. Using established criteria, the staff reviews plant-specific summary reports and determines whether a preimplementation audit is necessary. After completing its review, the staff issues a safety evaluation report documenting the acceptance of the licensee's proposals.

Since the issuance of Supplement 1 to NUREG-0737, no new requirements have been identified for completing the DCRDR. Through periodic resident and regional inspections, and 10 CFR 50.59 reviews, the staff will ensure that future modifications to plant control rooms and remote shutdown facilities are implemented in a manner consistent with the plant's approved DCRDR process and NRC acceptance criteria. The staff will review and evaluate advances in technology that may affect the design of the plant control room or remote shutdown facility through periodic plant inspections and by sponsoring research in advanced control room design. If changes to the current requirements are needed, they could be implemented using existing regulatory programs, as necessary, to ensure continued public health and safety.

18.2.4 Conclusions

The staff review at the time of the initial licensing of a facility determined that the information provided by the applicant was sufficient for the staff to conclude that Control Room Design met the staff acceptance criteria and the intent of all applicable regulations. The staff has and will continue to obtain new information related to this subject area through a variety of sources such as updates to the FSAR, research reports, operating plant events and routine plant inspections. The staff reviews this information and, in the past, has required licensees to take actions to upgrade the plant to provide continuing assurance of adequate protection of the public health and safety. For example, the requirements for the DCRDRs were established in response to the TMI accident and are contained in NUREG-0737, Supplement 1. The staff reviews and approves plant-specific DCRDR efforts and documents these approvals in published safety evaluation reports after the review is completed. Resident and regional inspections, and 10 CFR 50.59 reviews, will ensure that future modifications to plant control rooms and remote shutdown facilities are made in accordance with NRC approved DCRDR programs and NRC acceptance criteria. In addition, the staff will continue to review new information in this subject are and if the staff determines that new or different requirements are needed, the staff has the capability within the existing regulatory process to require additional analyses or plant modifications, as necessary, to ensure the continued health and safety of the public. In conclusion, the Commission concludes that the current regulatory process has and will continue to provide reasonable assurance that the licensing bases of all currently operating plants are sufficient to assure that operation is not inimical to the public health and safety.

18.3 Safety Parameter Display System

18.3.1 Scope

In addition to upgrading the design of their control rooms, licensees are to install a safety parameter display system (SPDS) as an aid to operating personnel in rapidly and reliably determining the safety status of the plant and in assessing whether abnormal conditions warrant corrective actions to avoid a degraded core.

18.3.2 Safety Issues and Regulatory Requirements

The safety issue addressed is to confirm that the design and implementation of the plant's SPDS facilitate the user's ability to rapidly and reliably determine the safety status of the plant.

In May 1980, requirements for commercial nuclear power plant licensees to install an SPDS were established as a result of the TMI accident. The basis for regulating the design and implementation of the plant's SPDS is given in the enclosure to Generic Letter 82-33, "Supplement 1 to NUREG-0737--Requirements for Emergency Response Capability."

18.3.3 Evolution of Current Licensing Basis

NUREG-0660, Item I.D.2, "Plant Safety Parameter Display Console," stated that "In conjunction with the control room design upgrade described in Item I.D.1, NRR will require all licensees and applicants to install a safety parameter display system that will display to operating personnel a minimum set of parameters (safety state vector) which define the safety status of the plant." In November 1980, the NRC published NUREG-0737, "Clarification of TMI Action Plan Items," which identified the specific requirements associated with the SPDS. Guidance published as NUREG-0695, "Functional Criteria for Emergency Response Facilities" (1980), and NUREG-0835, "Human Factors Acceptance Criteria for the Safety Parameter Display System, Draft Report for Comment" (1981), were also issued to the industry. In December 1982, "Supplement 1 to NUREG-0737--Requirements for Emergency Response Capability," was issued as Generic Letter 82-33. This document implemented the requirements to install an SPDS and provided additional clarification. For some, but not all, plants, the NRC issued confirmatory orders, which required plants to submit schedules for the design, installation, and implementation of an SPDS.

In 1986, the staff issued NUREG/CR-4797, "Progress Reviews of Six Safety Parameter Display Systems," and concluded that utilities may be having major difficulties in designing and implementing their SPDSs. The staff subsequently issued NRC Office of Inspection and Enforcement (IE) Information Notice (IN) 86-10, "Safety Parameter Display System Malfunctions" (1986), to inform licensees of the results of the survey. After issuing IN 86-10, the staff received several requests from licensees for extensions to implementation schedules, requests for clarification regarding the definition of an "operational SPDS," and questions about SPDS deficiencies and how to resolve them. In response to the continuing concerns related to SPDS designs, NRC published NUREG-1342, "A Status Report Regarding Industry Implementation of Safety Parameter Display Systems," which described methods used by some licensees to implement the SPDS in a manner acceptable to the staff and issued Generic Letter 89-06 in April 1989 requesting that licensees certify the operational status of their SPDSs to the NRC, using guidance contained in the generic letter and NUREG-1342. The staff is presently reviewing licensee submittals requested by GL 89-06.

Since the issuance of NUREG-0737, Supplement 1, and the guidance described above, no new requirements have been identified for installing and implementing the SPDS. Through the process of periodic resident and regional inspections and reviews, the staff will ensure that future modifications to plant SPDSs are implemented in a manner consistent with the plant's approved design and NRC acceptance criteria. The staff will review and evaluate advances in technology that may affect the design of the SPDS through periodic plant inspections and by sponsoring research in advanced SPDS design. If changes to the current requirements are needed, they could be implemented by rulemaking on existing requirements or under the backfit rule to ensure continued health and safety to the public.

18.3.4 Conclusions

The requirements for the SPDS were established in response to the TMI accident and are contained in NUREG-0737, Supplement 1. Since NUREG-0737, Supplement 1, was issued, there have been no new requirements for SPDS. The staff reviews and approves plant SPDS designs, and a safety evaluation report is issued after the staff completes its review. Resident and regional inspections and 10 CFR 50.59 reviews will ensure that future modifications to plant SPDS designs are made in accordance with NRC guidance and acceptance criteria. Advances in design technology will be reviewed by the staff and if changes to the existing requirements are necessary, they will be implemented within existing regulatory programs, as necessary, to ensure the continued public health and safety. In conclusion, the Commission concludes that the current regulatory process has and will continue to provide reasonable assurance that the licensing bases of all currently operating plants are sufficient to assure that operation is not inimical to the public health and safety.

19 RESOLUTION OF SAFETY ISSUES: TECHNICAL AND IMPLEMENTATION STATUS

19.1 Scope

The NRC has an integrated program in place for reviewing and analyzing operating experience in order to identify specific events and generic situations in which the margin of safety established by design through the licensing process has been degraded, or where new information or insights lead to new concerns. The program also includes steps to identify and implement corrective actions that will restore the intended margin of safety.

19.2 Safety Issues and Regulatory Requirements

NRC licensees must report any unexpected occurrence in operation that has actual or potential safety significance. Some events must be reported within one hour via dedicated direct phone lines, and many must be reported in writing within a few weeks. These written reports, required by 10 CFR 50.73, are called licensee event reports (LERs) and provide a clear, narrative description of the event and the cause of each component or system failure, if the cause is known. The staff reviews these LERs to determine the adequacy of short-term corrective actions and the need for possible action at other plants, or to identify potential generic problems and significant safety concerns warranting further study.

19.3 Regulatory Process and Implementation Status

For many safety-related operational events, NRC resident inspectors perform the initial NRC investigations, and the appropriate NRC regional office conducts reviews. In addition, the technical aspects of potentially significant operational events are studied by appropriate organizations within the NRC, including the Office for Analysis and Evaluation of Operational Data (AEOD) and the Offices of Nuclear Reactor Regulation (NRR) and Nuclear Regulatory Research (RES).

The AEOD analyzes and evaluates all operational safety data and provides a strong technical capability that is independent of regulatory activities associated with licensing and inspection. Engineering evaluations are performed to examine the implications of operating experience and to determine if intensive analysis and evaluation as a case study are warranted. If necessary, in-depth case studies are performed to determine the level of safety concern, and findings and recommendations are sent to the appropriate NRC office for action.

The AEOD recommendations and suggestions addressed to NRR are reviewed and prioritized according to a judgment of their safety significance. If an item appears to have a high degree of safety significance, the need for an information notice, generic letter, bulletin, or other appropriate prompt action is determined. If the recommendation does not appear to warrant immediate action, it is considered within NRR for appropriate action or a determination whether it can be addressed as part of an existing issue (such as a generic issue) or by creation of a new generic issue. If this occurs, the issue is formally transmitted to RES for its consideration and prioritization or for inclusion into an existing generic issue.

The AEOD also screens the recommendations and suggestions contained in its studies and evaluations for identification as potential generic issues. A generic issue is an issue that is applicable to all, several, or a class of reactors or reactor-related facilities. Such issues are identified to RES, which then evaluates and prioritizes the issue in accordance with established procedures (see below). Generic issues may also be suggested by individuals within the NRC, the Advisory Committee on Reactor Safeguards (ACRS), the nuclear power industry, or the public.

The generic issues management program comprises six distinct stages. In addition to the identification stage (discussed above), the stages are prioritization, resolution, imposition, implementation, and verification. Each new generic safety issue (GSI) is prioritized by developing a quantitative assessment of safety benefits (risk reduction) and impact (cost) for the utility, the NRC, and any other entities involved, as described in NUREG-0933, "A Prioritization of Generic Safety Issues." On the basis of the extent of potential risk reduction to the public and the value/impact ratio developed from this assessment, and as further adjusted by qualitative judgments, a priority is assigned to each GSI. Following peer review of the initial prioritization, a final priority is recommended and assigned.

Issues that receive a high or medium priority are designated for resolution by the staff. An issue given a low or drop priority is, by nature of the rating standard, of so low a public risk reduction potential that resolution of the issue is not pursued. All issues are documented in the catalog of generic issues maintained in NUREG-0933.

The resolution process requires the development of a plan and schedule for the work that needs to be done to resolve the issue. The plan also identifies needed resources and coordination points. Following completion of the technical studies, a final resolution package is prepared that includes a regulatory analysis describing various potential solutions and justification for any proposed requirements based on a consideration of value and impact. The resolution package is considered by the ACRS and by the Committee for Review of Generic Requirements (CRGR) if new requirements are proposed. Resolved issues are forwarded to NRR for imposition, implementation, and verification. This includes issuance of generic correspondence to licensees informing them of the issue resolution, establishment of an acceptable schedule for implementation of the resolution by the affected licensees, and verification that the required improvements have been made in an acceptable manner.

Value impact analyses were employed as part of the basis of resolving some GSIs. In the tradeoffs between net safety benefit and net cost, the remaining plant operating term ordinarily enters the calculations. Both the safety value and the cost impact can increase over time more than the cost impact, as would be the case when costs are largely one-time initial costs but the risk reduction benefit accumulates year after year with continued operation. Consequently, the consideration of extended plant life, and also any increase in population around nuclear plant sites, may alter the resolution bases of GSIs that have been resolved but not backfitted.

The staff performed a systematic evaluation of all GSIs resolved through October 1990 NUREG/CR 5382 to determine those whose resolution bases could be affected by an additional 20 years of plant life. A screening analysis was performed on 249

GSIs that were resolved through October 1990, and 139 GSIs were identified that did not result in backfit requirements. This total includes issues that were resolved without backfit and issues that were prioritized low and not considered further. Three GSIs were identified in which the value-impact estimates played a relatively significant role in the resolution and the revised estimates were judged to warrant a further reconsideration of decision not to backfit. These issues are:

- GSI-III.A.1.3(2) Maintain Supplies of Thyroid Blocking Agent
- GSI-82   Beyond-Design-Basis Accident in Fuel Pools
- GSI-101   BWR Water Level Redundancy

Due to considerations aside from license renewal, GSI-III.A.1.3(2) has been returned to the issue resolution process and is currently under reevaluation. The resolution bases of GSIs-82 and 101 were also reconsidered and the staff determined that the decision not to backfit is still appropriate.

The screening analysis of issues that were originally prioritized in the low category indicated that four issues could be placed in the medium category. These issues are:

- GSI-II.D.2   Research on Relief and Safety Valve Test Requirements
- GSI-III.D.2.1   Radiological Monitoring of Effluents
- GSI-35   Degradation of Internal Appurtenances in LWRs
- GSI-80   Pipe Break Effects on Control Rod Drive Hydraulic Lines in the Drywells of BWR Mark I and II Containments

The prioritization of the issues has been reconsidered and determined that GSIs 11.D.2, III.D.2.1, 35, and 80 should remain in the low category.

The generic issues management program was initiated in 1981. At that time, 511 issues were identified to be prioritized; 369 were TMI followup items (NUREGs-0660 and -0737) and 142 were identified by previous assessments of generic issues (NUREGs-0371 and -0471). These issues included 22 issues (Appendix A) that had previously been identified as unresolved safety issues (USIs). In the past 10 years, an additional 264 issues have been identified, for a total of 775; this number includes various human factors issues and issues identified by the staff assessment of the Chernobyl accident. As of December 1990, 721 issues have been resolved, including all the USIs. Of the remaining 54 issues, 26 are to be prioritized and 28 are in the resolution process.

The implementation status of USIs was recently reviewed by the NRR staff. NRR's findings indicate that, in general, most USIs have been implemented and that unimplemented USIs are being addressed on a schedule satisfactory to the staff. The implementation status of the remaining generic safety issues is currently being assessed by NRR.

19.4 Conclusions

The NRC has an effective program in place for reviewing and analyzing operating experience and other new information, and for implementing any necessary modifications at operating reactors. The process allows for early notification of licensees of potential concerns, if deemed necessary, or for more thorough evaluation through the generic issues management program. Plant modifications are implemented following an evaluation of various reasonable alternative solutions and justification based on an assessment of value and impact. The licensing basis for individual operating plants includes changes resulting from resolution of generic issues determined to be applicable.

APPENDIX A

UNRESOLVED SAFETY ISSUES FOR WHICH A FINAL
TECHNICAL RESOLUTION HAS BEEN ACHIEVED

Number	Title	Report Number	Date
A-1	Water Hammer	NUREG-0927, Rev. 1 NUREG-0933	March 1984
A-2	Asymmetric Blowdown Loads on Reactor Primary Coolant Systems	NUREG-0609	November 1980
A-3	Westinghouse Steam Generator Tube Integrity	NUREG-0844	September 1988
A-4	CE Steam Generator Tube Integrity	NUREG-0844	September 1988
A-5	B&W Steam Generator Tube Integrity	NUREG-0844	September 1988
A-6	Mark I Short-Term Program	NUREG-0408	December 1977
A-7	Mark I Long-Term Program NUREG-0661 Suppl.	NUREG-0661	July 1980
A-8	Mark II Containment Pool Dynamic Loads	NUREG-0808	August 1981
A-9	Anticipated Transients Without Scram	NUREG-0460, Vol. 4	September 1980
A-10	BWR Feedwater Nozzle Cracking	NUREG-0619	November 1980
A-11	Reactor Vessel Material Toughness	NUREG-0744, Rev. 1	October 1982
A-12	Fracture Toughness of Steam Generator and Reactor Coolant Pump Supports	NUREG-0577, Rev. 1	September 1982

UNRESOLVED SAFETY ISSUES FOR WHICH A FINAL
TECHNICAL RESOLUTION HAS BEEN ACHIEVED
(continued)

Number	Title	Report Number	Date
A-17	Systems Interactions	NUREG-1229 Generic Letter 89-18	August 1989
A-24	Qualification of Class IE Safety-Related Equipment	NUREG-0588, Rev. 1	July 1981
A-26	Reactor Vessel Pressure Transient Protection	NUREG-0224	September 1978
A-31	Residual Heat Removal Shutdown Requirements	SRP 5.4.7	1978
A-36	Control of Heavy Loads Near Spent Fuel	NUREG-0612	July 1980
A-39	Determination of SRV Pool Dynamic Loads and Pressure Transients	NUREG-0802	September 1982
A-40	Seismic Design Criteria	NUREG-1233	September 1989
A-42	Pipe Cracks in Boiling Water Reactors	NUREG-0313, Rev. 1	July 1980
A-43	Containment Emergency Sump Performance	NUREG-0897, Rev. 1	October 1985
A-44	Station Blackout	Regulatory Guide 1.155 NUREG-1032 NUREG-1109	August 1988 June 1988 June 1988
A-45	Shutdown Decay Heat Removal Requirements	NUREG-1289 NUREG/CR-5230	September 1988
A-46	Seismic Qualification of Equipment in Operating Plants	NUREG-1030 NUREG-1211	February 1987
A-47	Safety Implications of Control Systems	NUREG-1217 NUREG-1218 Generic Letter 89-19	September 1989
A-48	Hydrogen Control Measures and Effects of Hydrogen Burns on Safety Equipment	NUREG-1370	September 1989
A-49	Pressurized Thermal Shock	Regulatory Guide 1.154	February 1987

NRC FORM 335
(2-89)
NRCM 1102,
3201, 3202

U.S. NUCLEAR REGULATORY COMMISSION

BIBLIOGRAPHIC DATA SHEET

(See instructions on the reverse)

1. REPORT NUMBER
(Assigned by NRC. Add Vol., Supp., Rev., and Addendum Numbers, if any.)

NUREG-1412

2. TITLE AND SUBTITLE

Foundation for the Adequacy of the Licensing Bases

A Supplement to the Statement of Considerations for the Rule on Nuclear Power Plant License Renewal (10 CFR Part 54) Final Report

3. DATE REPORT PUBLISHED

MONTH	YEAR
December	1991

4. FIN OR GRANT NUMBER

5. AUTHOR(S)

6. TYPE OF REPORT

Regulatory

7. PERIOD COVERED *(Inclusive Dates)*

8. PERFORMING ORGANIZATION – NAME AND ADDRESS *(If NRC, provide Division, Office or Region, U.S. Nuclear Regulatory Commission, and mailing address; if contractor, provide name and mailing address.)*

Office of Nuclear Reactor Regulation
U.S. Nuclear Regulatory Commission
Washington, DC 20555

9. SPONSORING ORGANIZATION – NAME AND ADDRESS *(If NRC, type "Same as above"; if contractor, provide NRC Division, Office or Region, U.S. Nuclear Regulatory Commission, and mailing address.)*

Same as above

10. SUPPLEMENTARY NOTES

11. ABSTRACT *(200 words or less)*

The objective of this report is to describe the regulatory processes that assures that any plant-specific licensing bases will provide reasonable assurance that the operation of nuclear power plants will not be inimical to the public health and safety to the end of the renewal period. It is on the adequacy of this process that the Commission has determined that a formal renewal licensing review against the full range of current safety requirements would not add significantly to safety and is not needed to assure that continued operation throughout the renewal term is not inimical to the public health and safety or common defense and security. This document illustrates in general terms how the regulatory process has evolved in major safety issue areas. It also provides examples illustrating why it is unnecessary to re-review an operating plant's basis, except for age-related degradation unique to license renewal, at the time of license renewal. The report is a supplement to the Statement of Considerations for the Nuclear Regulatory Commission's rule (10 CFR Part 54) that establishes the criteria and standards governing nuclear power plant license renewal.

12. KEY WORDS/DESCRIPTORS *(List words or phrases that will assist researchers in locating the report.)*

Licensing Bases, License Renewal, Nuclear Power Plant

13. AVAILABILITY STATEMENT

Unlimited

14. SECURITY CLASSIFICATION

(This Page)

Unclassified

(This Report)

Unclassified

15. NUMBER OF PAGES

16. PRICE

www.ingramcontent.com/pod-product-compliance
Lightning Source LLC
Chambersburg PA
CBHW081459170526
45166CB00008B/2487